EPISERVER

THE MISSING MANUAL

Richard Simonsbacka

Contents

About the book

I started writing this book because I think it's needed. There are many learning resources for EPiServer - documentation at EPISERVER WORLD, and Joel Abrahamson architecture and book, a wealth of blog posts and class room based training arranged by EPiServer

All of those are great, but I find few things are as valuable as a book when learning a technology. For someone that is completely new to the topic a book can provide a structured way of approaching the technology, at a custom pace, and for someone that already knows the topic a book is often a great way of discovering hidden gems in the technology and seeing someone else's, the authors, perspective on how to develop with it.

At least that's what I think. I'm also writing this book as it's fun to do and I hope to enrich my own knowledge about EPiServer using, administrating and development based on reader feedback. If you have an idea on how to make the book better or about a topic that you'd like it to cover don't hesitate to comment on zianneson.com or e-mail me at zianneson@zianneson.com.

How to read this book

The book mixes theory with practical hands-on tutorial. As such you will get most out of the book by reading it close to a computer and performing exercises/instructions on it. However, my intention is also that it should be possible to only read the book, coming back later to practical guides.

Architecture

This topic describes the architecture of the Episerver platform, with an introduction to the system foundation and related components and products.

Platform

Episerver consists of the Episerver framework with a core user interface, CMS and Commerce for content management and e-commerce, and other modules supporting marketing automation and social interaction features. The Episerver Add-on Store contains a broad selection of ready-to-use add-ons to extend the functionality of a solution, and you also can build your own custom add-ons.

Episerver has an open and layered architecture, allowing for almost any type of integration using standard technology.

EPISERVER
USER GUIDE

© Episerver
THE MISSING MANUAL

Introduction

The features and functionality of the entire Episerver platform are described in an online help that opens in a web browser. The online help covers CMS for content management, Commerce fore-commerce functionality, Find for extended search, and Episerver add-ons. It is either accessed from within the Episerver platform or from Episerver World. The online help is also divided into a number of PDFs for users who prefer those or want to print the documentation.

This PDF describes the features and functionality of Episerver CMS. PDFs for Episerver Commerce and Find can be found on Episerver World. The user documentation is intended for editors, administrators, marketers and merchandisers, working with tasks as described in Roles and tasks.

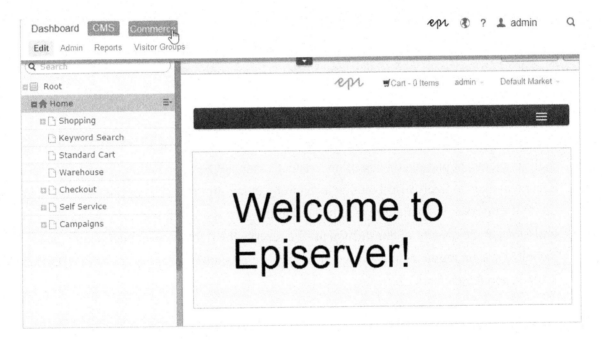

Developer guides and technical documentation are also found on Episerver World.

Features, licenses and releases

The user documentation is continuously updated and covers the latest releases for the Episerver platform.

CMS Episerver CMS is the core part of the Episerver platform providing advanced content creation and publishing features for all types of website content. CMS features are available in all Episerver installations.

Commerce Episerver Commerce adds complete e-commerce capabilities to the core functionality in CMS. Commerce requires additional license activation.

Addons Add-ons extend the Episerver capabilities with features like advanced search, multivariate testing, and social media integration. Some add-ons are free, others require license activation. Add-ons by Episerver are described in the online help.

  Due to frequent feature releases, this user guide may describe functionality that is not yet available on your website. Refer to What's new to find out in which area and release a specific feature became available.

Copyright notice

About Episerver

The base of the Episerver platform is the CMS (Content Management System) with its core features for online content creation, publishing, and website management. The platform can be extended with Episerver Commerce for managing e-commerce tasks, and Episerver Find for building advanced search features, as well as a broad selection of other add-ons from both Episerver and third-parties.

Features

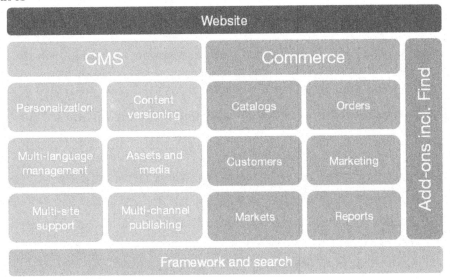

CMS

Episerver CMS is a powerful yet easy to use web content management platform, based on cutting edge technology. The intuitive user interface and superior usability of Episerver CMS allow both experienced and occasional users to efficiently manage website content. Refer to the Managing content section in the CMS Editor user guide for more information on how to work with CMS features.

Commerce

Adding Episerver Commerce to your CMS solution brings e-commerce functionality such as catalog, customer and order management, combining the powerful content publishing and display features of CMS with advanced back-end online store management. Refer to the Commerce user guide for more information about Episerver Commerce features.

Add-ons

There are many add-ons available for extending your Episerver solution. The add-ons from Episerver described in this documentation add features such as advanced search, Google Analytics and social media integrations. Refer to the Add-ons section in the online help for more information about Episerver add-ons.

Refer to Introduction and What 's new for information about licenses and recent features.

What's new?

The Episerver user guide describes features in the Episerver platform, including CMS for content management and Commerce fore-commerce management, and add-ons from Episerver. New features are continuously made available through Episerver updates.

This user guide (16-6) describes **features added up until and including update 127** for Episerver; see Episerver World for previous user guide versions.

Area	Features and updates
CMS	» Access rights was revised to expand its information. » You can watch the following demonstration video, Managing access rights. (6:39 minutes) » You can watch the following demonstration video, Publishing content. (4:18 minutes)
Commerce	» A new discount lets you give free shipping to customers who buy a minimum number of items. See Buy products for free shipping. (update 125) » A new discount lets you give free shipping to customers who spend a minimum amount on an order. See Spend for free shipping. (update 125) » A new discount lets you give a discount on every item to customers who order a minimum number of items. See Buy products, get discount on all selected. (update 126) » A new discount lets you give the most expensive item for free to customers who order a minimum number of items. See Get most expensive for free. (update 126) » You can watch the following Demo of creating a campaign and discount video. (4:43 minutes)
Addons	» Episerver Forms has a Marketing Automation Integration connector that lets you connect Episerver form fields to a Digital Experience Hub (DXH) connector database.

Getting started

This section describes how to log in to an Episerver website, access features and navigate the different views. Note that the login procedure may be different from what is described here, depending on how your website and infrastructure are set up. The examples described here are based on a "standard" installation of Episerver with sample templates.

Logging in

As an editor or administrator, you usually log in to your website using a specified URL, a login button or link. Enter your username and password in the Episerver login dialog, and click **Log In**.

Accessing features

What you are allowed to do after logging in depends on your implementation and your access rights, since these control the options you see. When logged in, the Episerver quick access menu is displayed in the upper right corner.

Selecting **CMS Edit** takes you to the edit view as well as other parts of the system. You can go directly to your personal dashboard by selecting the **Dashboard** option.

Navigation

Pull down the **global menu**, available at the very top, to navigate around. The menu displays the different products and systems integrated with your website. Select, for instance, **CMS** to display available options in the submenu.

Your Menu options vary depending on your access rights. The user guide examples assume that the user has full permissions to all functions in Episerver

Your Menu options vary depending on your access rights. The user guide examples assume that the user has full permissions to all functions in Episerver

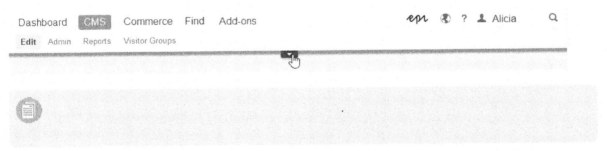

Your Menu options vary dependingon your access rights. Thse user guide examples assume

© Episerver

Next steps

Refer to the sections below for more information.

» User interface and Roles and tasks in the CMS Editor User Guide for information about the Episerver user interface and roles.

» Managing content in the CMS Editor User Guide for information on how to create and publish content.

» Administration interface in the CMS Administrator User Guide for information on how to administer and configure settings in Episerver.

» Commerce User Guide for information on how to work with e-commerce tasks, if you have Episerver Commerce installed.

» Find User Guide for information on how to work with search optimization, if you have Episerver Find installed.

» Add-ons section in the online help for information on how to use add-ons from Episerver, if you have any of these installed.

User interface

 The Episerver user interface is flexible allowing developers to plug-in customized features when implementing websites. This description refers to a "standard installation" of Episerver without customizations.

The **toolbar** and the **panes** in the Episerver edit view provide easy access to functions when working with content. When entering the edit view, you have the global menu and the toolbar at the top, and adjustable panes to the left and right.

The **global menu** provides access to other areas of the Episerver platform. You can also access this user guide, your user settings and a global search from the global menu. Note that you in some areas need to pull down the global menu with the arrow.

The toolbar contains features like preview and view options, and lets you add items such as pages or blocks.

The page information area below the toolbar displays the page name, the path to the page and when it was last auto saved. If it is possible to undo the latest changes you did to the content, there is also an **Undo** link.

A notification toolbar may be displayed below the toolbar showing confirmation and error messages and such.

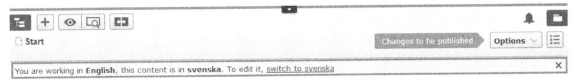

General features

» Context-sensitive actions. Some features are only available in certain context. The toolbar, for instance, presents relevant actions depending on what you are currently doing, and the add (+) button also offers context-sensitive options.

» Primary actions. Some actions open in a new window, for example, a page delete confirmation. The background is then dimmed, meaning that you must finish the primary action to continue.

» Action feedback and notifications. Successful actions are confirmed by a message in the notification bar. A notification may also appear in case of an error requiring you to take action. » Drag-and-drop operations. Drag-and-drop is supported in many areas. For instance, you can drag pages, media files and blocks into the rich-text area or content areas, or re-arrange the page tree structure using drag-and-drop.

» Tooltips. Hover the mouse over a button or field and a short tooltip is displayed. » Keyboard commands. Standard keyboard commands are supported in many areas, for instance, when moving pages in the page tree or in the rich-text editor.

» Search. Supported in many areas to locate, for instance, pages in the page tree or media in the folder structure.

» Adaptable work environment. Resize and pin the panes depending on what you are currently doing, and add and remove gadgets of your choice for quick access to functionality. » Support for time zones. Publishing actions in the edit view are done in your local time zone, whereas administrative actions are based on server time.

» Context menus are available in many areas, for instance, in the panes, the page tree and in item listings. The menu displays different available options depending on where in the interface you are and what you are doing.

Panes

The user interface has a left-hand and a right-hand pane, which can be adjusted and extended with additional gadgets.

Pane pin is used for expanding and locking the panes in an open position.

Settings fora pane or a gadget allow you to configure or remove a gadget, or rearrange gadgets in a pane.

Left-hand navigation pane

Contains the page tree structure (Pages), language branch (Sites) navigation, tasks management (Tasks), and project items (Project Items) by default.

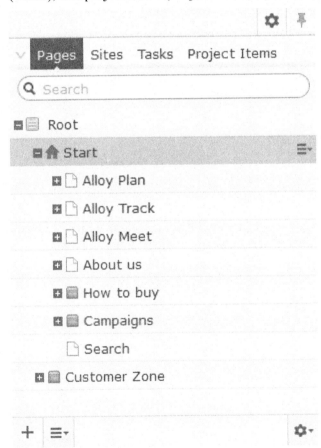

Right-hand assets pane

Contains the Media and Blocks folder structures by default. Drag and drop items from the assets pane into the content you are currently working on.

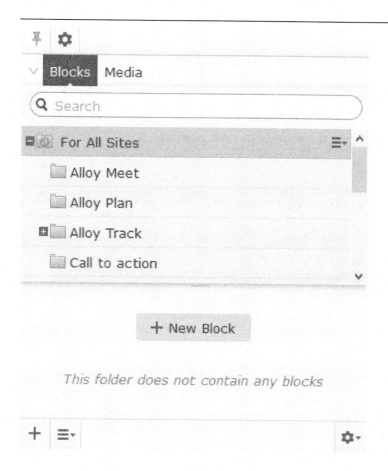

Editing

The toolbar on top displays an actions menu with context-sensitive publishing options which vary depending on content status and user access rights. During editing, content status and auto save information are displayed.

When working with content such as pages and blocks in CMS, or catalog content in Commerce, there are two editing views, **On-Page Editing** and **All Properties**, with toggle buttons to switch between them.

On-Page Editing provides quick access to direct editing of a selection of content properties.

All Properties provides access to all available properties including more advanced ones such as access rights and language settings.

When editing content properties, these options are common when adding, deleting, or selecting items:

Click to select, for instance, a category.

Click to select, for instance, an image in a media folder, or a page in the page tree.

Click to remove, for instance, a category.

Real-time Updates

Episerver is by default set up to display comments and changes to projects in the user interface immediately. If Episerver experiences problems with the these immediate updates, the following dialog box is displayed.

Real-time Updates

A real-time connection could not be established with the
server. This may be caused by incorrect configuration. Please
read the documentation for further information.

OK

The *Real-time Updates* dialog box can have two main reasons. It appears either because Episerver has encountered a network problem and cannot connect to the web server or because your system does not have the Web Socket protocol enabled, which is used for the real-time updates of the user interface.

Network problems

Network problems occur if there is an error in the web server configuration, or if there are problems with your corporate network or the with Internet connection. As long as there is a problem, you are unable to continue working with Episerver.

Whatever the reason, Episerver will try to reconnect to the web server. If it still cannot connect after a number of attempts, you get an error message saying: "The server has been unavailable for an extended period of time. Please verify the internet connection and refresh the browser."

Contact your IT department or Internet service provider if the problem persists.

Web Socket support

Web Socket is an Internet protocol used to automatically update the Episerver user interface.

If you are using the projects feature, the Web Socket protocol is used to check for new or updated comments and project items from other users. As soon as one of your colleagues adds a comment or project item, Episerver automatically updates your user interface and displays the comment or item (assuming you have the projects interface open).

However, the Web Socket protocol must have been enabled for your system by an administrator for the real-time updates to work. If it has not been enabled, you can still work with Episerver and with projects and comments but you need to manually refresh the user interface with the **Refresh** button to see new comments or items.

Depending on the system configuration when the Web Socket protocol is disabled, you may or may not see the *Real-time Updates* dialog box.

My settings

Under the global menu > *your user name* > **my settings**, you can change some of your account settings. Username and password are usually set in an external system, such as Windows or an SQL database, and cannot be changed in Episerver in that case. In the **Display Options** tab, you find the following settings:

>> Personal language. Select your desired user interface language from the drop-down. This settings affects the language of the user interface, such as texts in buttons and dialog boxes. It does not affect the language of your website.

>> Limit touch support. On touch-screen devices, browsers may have difficulties prioritizing between input from the screen and from the mouse, which may cause problems such as making it impossible to resize panes using the mouse. If you experience problems with Episerver and the touch screen, enable the Limit touch support feature. This feature prioritizes mouse input over touch screen input and enables the mouse for interactions such as drag and drop, resizing of panes, and so on.

>> **Reset all views to system default**. You can reset your user interface views to the settings. If you have changes to the user interfaces, such as added, moved or deleted gadgets, these changes are undone and the default views are displayed.

Roles and tasks

Episerver is designed for interaction with website visitors, as well as collaboration between users. A user in Episerver is someone working with different parts of the platform. A user can belong to one or more user groups and roles, depending on their tasks as well as the size and setup of the organization.

Typical roles and related tasks are described below. Refer to Setting access rights in the CMS Administrator User Guide for information on how to configure user groups and roles in Episerver.

Visitor

A visitor is someone who visits the website to find information or to use available services, on an ecommerce website possibly with purchasing intentions. Purchasing on an e-commerce website can be done either "anonymously" (payment and shipping details provided), or by registering an account. Visitors may also contribute to website content as community members, which usually requires registration of an account profile.

Community member

Content may be added by visitors or community members, if social features and community functionality are available for the website. This content includes forum and blog postings, reviews, ratings and comments, in which case there might be a need for monitoring this type of content on the website. Monitoring can be done for instance by an editor, or a specific moderator role for large websites and online communities.

Content editor

A content editor is someone with access to the editorial interface who creates and publishes content on the website. Content editors with good knowledge of the website content work with search optimization for selected content in search results. Editors may also want to follow-up on content with unusually high or low conversion rate in order to update or delete this content.

Marketer

A marketer creates content and campaigns with targeted banner advertisements to ensure customers have consistent on site experience of the various marketing channels. Furthermore, the marketer monitors campaign KPIs to optimize page conversion. A marketer with good knowledge of the website content may also want to monitor search statistics in order to optimize search for campaigns and promote content.

Merchandiser

A merchandiser typically works with stock on an e-commerce website to ensure that the strongest products are put in focus. This role also creates landing pages, sets product pricing, coordinates cross product selling, oversees delivery and distribution of stock, and deals with suppliers. This user wants to be able to identify search queries with unusually high or low conversion rates, in order to adjust either the search or the product line.

Website owner

A website owner is someone with overall responsibility for the content and performance of one or more websites. This user monitors website activities such as page conversions, customer reviews or sales progress. Rarely creates content but can be involved in the approval of content created by others. A website owner may have administrative access rights and may be able to install selected add-ons on the website.

Administrator

An administrator works with configuration of various system settings from the administration user interface, including search, languages, user access rights and visitor groups for personalized content. Administrators may also install add-ons on the website. Administrators usually have extended access rights compared to other user groups, and can access all parts of the Episerver platform.

Developer

A developer is someone with programming skills working with the setup and implementation of the website, as well as maintenance and development of new functionality. Creates the content templates for pages, blocks and catalog content used by editors in CMS and Commerce, configures e-commerce settings, and manages the index and customized search features in Find. Developers may also install add-ons on the website.

CMS **Finding content**

Follow these tips to find Episerver content within edit view.

What you know	Action
Content location within page, asset or block structure	Use the page tree structure or the blocks and media assets pane and select content.
Content location on the website	Use the preview option, navigate to the content on the website.
Content is associated with a project	Use the projects gadget or, if you are using projects feature, the project overview or the project items navigation pane.
Text within content title or body	Use either the global search in the global menu or the search fields at the top of the navigation and assets panes. >> The global search can search all types of content on the website (depending on the configured search provider).

What you know	Action
	≫ Search fields in navigation and assets panes search for content in respective pane and related dialogs. See Search for search tips. 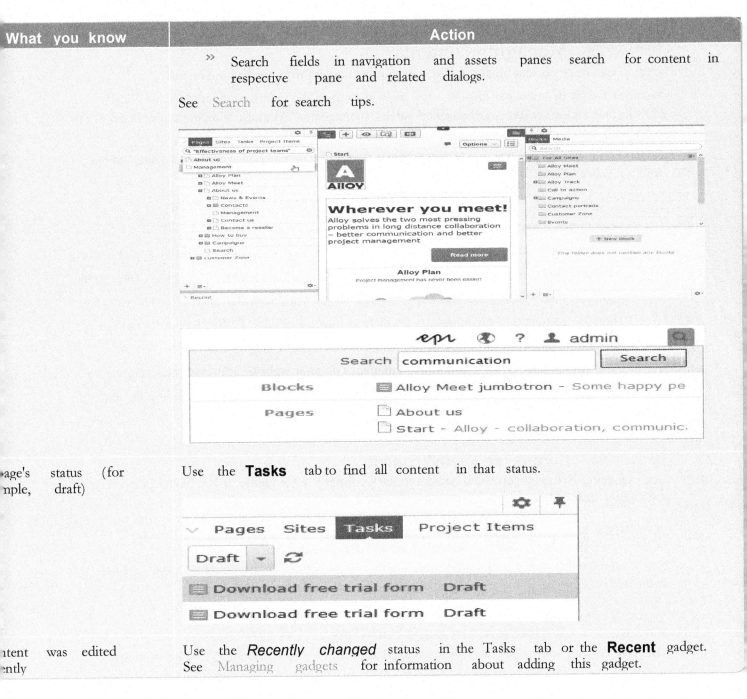
age's status (for mple, draft)	Use the **Tasks** tab to find all content in that status.
tent was edited ntly	Use the *Recently changed* status in the Tasks tab or the **Recent** gadget. See Managing gadgets for information about adding this gadget.

THE MISSING MANUAL

CMS Managing content

Content can be pages and blocks in CMS, or product content from the catalog on an

E-commerce site.Content can also be assets such as images and videos, or documents in Word or PDF format. Episerver has a sophisticated version management features, allowing multiple editors to work with draft versions, before approving and publishing the content.

Content on a website can originate from different sources, depending on where on the site and by whom it was created.

>> Editors and marketers, or merchandisers can create content internally, on an
  e-commerce website.

>> A visitor community member can create content externally through interactive social features
  on the website, if these are available.You can preview draft content before publishing, so that
  you can verify the content before publishing. When working with personalization, you can
  preview content the way it appears for different visitor groups. To further limit access to
  content that is work-in-progress, you also can set access rights for content from the edit
  view.If you have content in multiple languages on your website, Episerver has advanced
  features for managing translation of content into additional languages, including the use of
  fallback and replacement languages.

Commerce Commerce-related content

See Managing e-commerce-related content in the Commerce User Guide, if you have Episerver Commerce installed.

Addons Optimizing content to improve search

See Working with content to optimize search in the Find User Guide, if you have Episerver Find installed.

CMS Working in On-page edit view

On-page editing is what you see when accessing the edit view in Episerver. From here you can instantly start editing content; the areas that are available for editing are highlighted. The available areas depend on how the page type has been defined and each area is set up with a property type which controls what you can do with each area. One area can, for example, have an image property type and is intended for images; another area can have a page description property and is intended for text describing the page.

This means that you will have different editing options when clicking an area, depending on the area's type of property typically for a page, you can edit page name, description, the main editorial area, and perhaps a content area with blocks.

Areas and properties available for editing are implementation-specific, and depend on the type of property and how the rendering is built on your website.

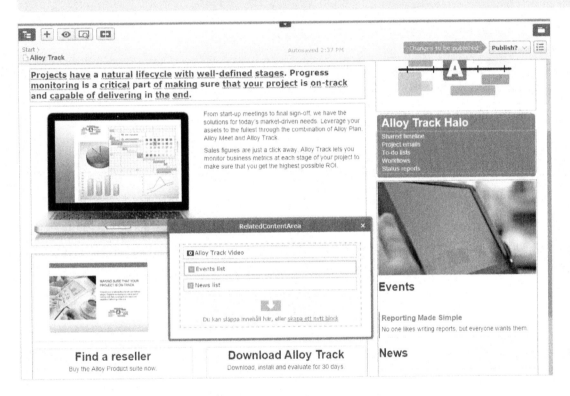

EXAMPLE: Editing a page through on-page edit

Here we describe how to edit a page, but the procedure is similar when editing, for instance, blocks or catalog content if you have Commerce installed.

1. Select the page to edit from the page tree in the navigation pane.
2. Click an area to edit (property names will be displayed on mouse-over).
3. Make your changes by updating the content properties as needed. Refer to using the rich-text editor for information on how to add and edit text in this type of property.

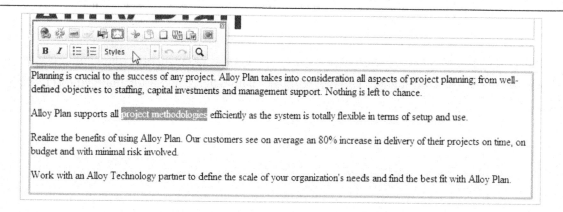

4. Your changes are automatically saved, and a draft version of the content (page, block etc) is created.

5. At any time, you can access the preview option at the top to see what the published version will look like.

6. When done, publish the content or apply any of the other options described in Publishing and managing versions.

Editing additional properties

The on-page editing view contains a number of properties which are reached by scrolling to the very top of the page with the wheel on your mouse. These are called basic info properties and can be used to add a simple address, set access rights for a page or to change the name in the URL and so on.

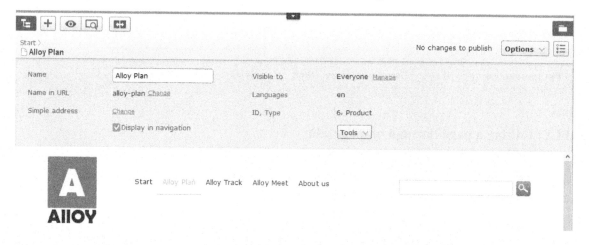

Not all properties are available in the on-page editing view; to see all properties, switch to the All Properties editing view.

in the toolbar to access these additional properties.

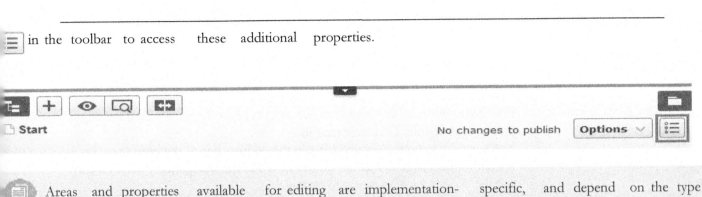

Areas and properties available for editing are implementation- specific, and depend on the type of content on your website. This topic describes some tabs and properties as they appear in the Episerver sample templates.

CMS Working in All properties editing view

The **All properties** editing view provides editing access to all properties available for content, including those that are not available in the On-page editing view. Select **All Properties**

Global properties
If you have content in multiple languages on your website, some properties may be locked for editing in a specific language. These properties are "globally shared" and you can edit them only in the master language. See Translating content.

Basic info properties
The top gray area displays the basic informational properties which are also accessible in the on-page editing view.

Simple address   This is a unique URL that can be added to frequently requested pages

On your website, letting visitors locate the page simply by typing the simple address name directly after the main URL of the website. Providing, for example, *products* as a simple address lets you find the page just by entering the URL (for example, *http://www.company.com/products*)in the address bar of the browser even if the Products page is located further down in the navigation tree. You can use the Simple address report to manage

Tabs

All other properties are organized using **tabs**. These can be added and modified in code and from the administrative interface.

Content tab

The Content tab contains properties for entering the main content, when editing a page or a block in CMS, or catalog content if you have Episerver Commerce installed.

Settings tab

The **Settings** tab is default and contains a set of built-in properties for managing publication dates, sort order and shortcuts.

CMS **Creating content**

Content can be pages or blocks in Episerver CMS, or catalog entries in Episerver Commerce. This topic describes creating types of content in Episerver.

Creating a page

You create a page from **page types** that contain the properties where information is added.

1. In the page tree structure, select the page under which you want to add the new page.
2. Select New Page from the context menu or the add button on the toolbar.
3. Select a page type from the list of available page types, and provide a name for the page.
4. Add information in the various properties available for editing, such as the rich-text editor if this is part of the selected page type.
5. Preview the page before sending it for approval or publishing it.

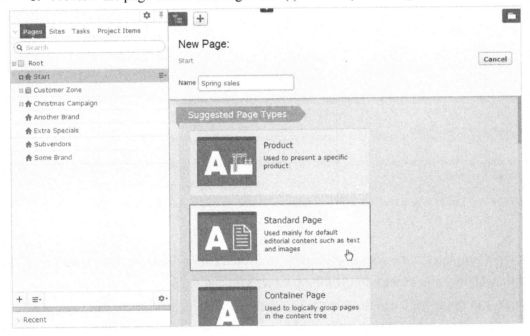

Creating a block

You create a block in a similar fashion as a page.

1. Select Create new block from the Blocks tab in the assets pane, or from the add button on the toolbar, in which case you are prompted to define a location for the new block.
2. Select a block type from the list of available blocks.
3. Provide a name for the block.

4. Add information for the block

You also can create a block directly from content areas in pages.

On an e-commerce website, content can be catalog entries such as products and variants if you have Episerver Commerce installed. The creation of this type of content is similar to the creation of pages and blocks in CMS. See Managing content in the Commerce User Guide.

CMS Editing content using the rich-text editor

The **TinyMCE rich-text editor** is a property where you can enter information such as text with formatting, images, tables and links to other content. The rich-text editor is used for both pages and blocks in Episerver CMS, and for catalog content if you have Episerver Commerce installed.

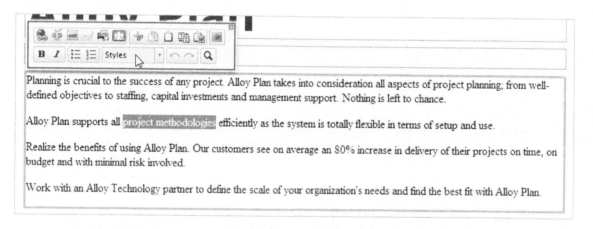

The Episerver sample templates come with a selection of activated editor functions (buttons). You can activate additional buttons from the administration view. You can drag and drop blocks into the editor area, and pages from the page tree to create links. Spell checking is available either from the browser you are using, or through the Spell checker add-on from Episerver.

Formatting

The **Styles** option displays a drop-down list with predefined style formats retrieved from the CSS style sheets on the website, to be used when formatting text.

You can extend the styles list to include specific formatting for an introduction text, a predefined table layout, or to position images with floating text in the editor area.

Copying and pasting

When you copy and paste text from external sources, you want to avoid including undesired formatting tags. You should work with plain text, or only use the **Copy from Word** option when copying from properly formatted Word documents.

Toggle paste as rich or plain text pastes the text as rich text by default; toggle to paste as plain text. Use keys Ctrl+v or Cmd+v to paste the text. Then use predefined styles to format the text as desired.

Paste from Word keeps the formatting from the Word document. Use keys Ctrl+v or Cmd+v to paste the content into the Paste from Word window, and click **Insert** to insert the content into the page.

To transform the text formatting from Word into the website's style, the headings and body text must be formatted using available template styles in Word. When you copy and paste text from Word, a "heading 2"in Word is converted into the"heading 2"using the website styles.

Keyboard shortcuts

The following standard keyboard shortcuts are supported in the rich-text editor:

Command	Shortcut keys
Select all	Ctrl+a or Cmd+a
Undo	Ctrl+z or Cmd+z
Redo	Ctrl+y or Cmd+z
Bold	Ctrl+b or Cmd+b
Italic	Ctrl+i or Cmd+i
Underline	Ctrl+u or Cmd+u
Copy	Ctrl+c or Cmd+c
Cut	Ctrl+x or Cmd+x
Paste	Ctrl+v or Cmd+v
H1–H6 headings	Ctrl+1–6 or Cmd+1–6
Paragraph break	Enter or Control+o
Line break	Shift+Enter

 Depending on customizations and the browser you are using, certain shortcuts may not work as described.

CMS Adding and editing images

A web page does not embed images, instead it links to the media library where images are stored. To display images in content, the images must be available in a folder in the **Media** structure. You can edit images inside Episerver using the **Image Editor**, providing basic image editing features such as cropping, resizing and transforming.

Adding images to content

Images often are added to content through an **image link property**, where you simply select an image from a folder under the **Media** tab, and the image is automatically placed and displayed in the content.

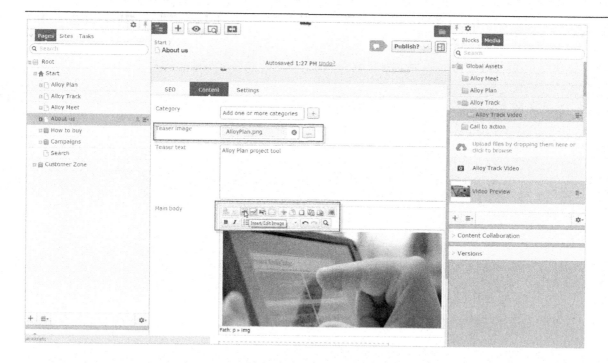

You also can add images to the **rich-text editor** or **content areas** in pages or blocks, either through **drag-and-drop** directly from **Media**, orby using the **toolbar** in the rich-text editor:

1. Place the cursor in the editor area where you the image.
2. Click the Insert/edit image button on the editor toolbar.
3. Enter an Image description for the image. The image description is important when a reader has turned off the display of images in the browser or when a visually impaired user is using a screen reader.
4. Type a Title for the image. The title is shown when the reader moves a mouse over the image.
5. Select the desired image in the Media folder structure.
6. Click Insert.

Editing images

Do one of the following to access the **Image Editor** options:

>> In the rich-text editor, click the image and then select Image Editor in the toolbar.

>> Locate the desired image in Media, and select Open in Image Editor from the context menu.

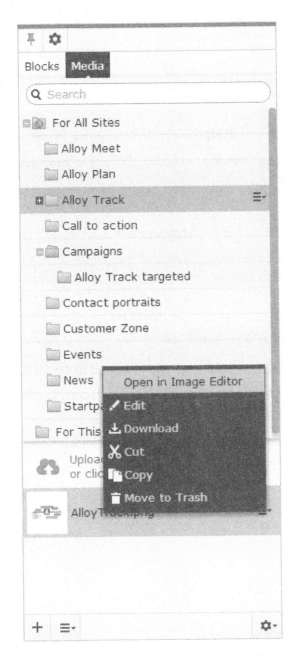

Image editing features include cropping, resizing and transforming:

Image editing	Description
Crop	Enter values for **Top**, **Left**, **Width** and **Height**, or draw a selection in the

Image editing	Description
	image to crop. **Note:** Remember to click **Apply** to save the changes before proceeding.
Resize	Enter values for **Width** and **Height**, or move a corner handle to resize, keep **Constrain proportions** selected to retain the image proportions. **Note:** Remember to click **Apply** to save the changes before proceeding.
Transform	Flip or rotate the image, or select **Grayscale** to convert to grayscale.
Preset values	Apply preset values for cropping and resizing, if such values are configured for the website.

You can save an edited image file as a copy, or replace the original file:

>> Save as a copy. If the image was selected in a page or block, the copy is saved in the local folder. If the edited image was selected in the media structure, the (renamed) copy is saved in the same folder as the original.

>> Replace original image. This action affects places on the website where the image is used.

the **Edit** option in the context menu for an image allows you to edit the **metadata**, which for an image can be photographer, description and copyright information and such. See Media.

Removing images from content

For an image in:

>> an image property , click the remove option.
>> a content area , select Remove in the context menu.
>> the rich- text editor , click the image and delete it.

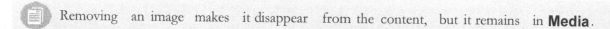
Removing an image makes it disappear from the content, but it remains in **Media**.

Changing image properties

You can control the display of images in the **rich-text editor** through a set of properties:

1. Select the image you want to change in the editor area

2. Click the Insert/edit image button on the editor toolbar.
3. On the General tab, you can change the Title and Image description. Click Update to save your changes.
4. On the Appearance tab, you have the following options of positioning images in relation to text.

   Click Update to save your changes.

CMS Adding links

Links are used on websites to link to content. In Episerver CMS, the following link types are available by default.

» Page. Links from one page to another on the same website.

» Media. Links to images, documents and other media files stored on the web server. »

email. Links to create an email message with the linked email address entered.

» External. Links to content on other websites or media on file shares.

» Anchor. Links to sections within a page, allowing readers to jump between topics on a page.

You also can use **shortcuts**, a specific type of link used for navigation and reusing existing website information. See All Properties editing view.

Creating a Link

You can create a link in the rich-text editor through drag-and-drop (pages and media files), or by selecting text and clicking the **insert/edit link** button in the toolbar, which displays the

Link dialog box.

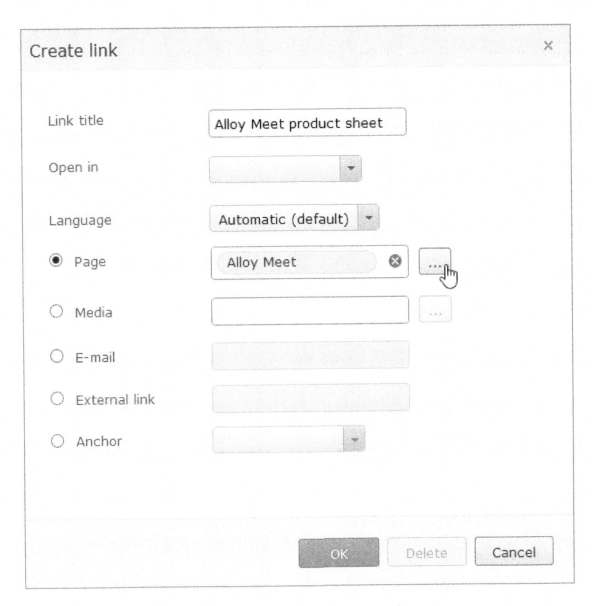

>> Link title displays as descriptive text for the link, such as on mouse-over.

>> Open in lets you display the link in a new window, often used for links to external sources.

>> Language lets you link to a specific language version for content. Automatic directs visitors to the detected browsing language version.

Options for linking to a page on the same website

>> Place the cursor in the rich-text editor where you want the link, and drag the desired page from the page tree into the location. The page name becomes the link name (which you can edit, if needed).

>> Select the text where you want the link in the rich-text editor, click insert/edit link, and use the Page option to select a page to link to.

Linking to a page on an external website

Select the text where you want the link in the rich-text editor, click **insert/edit link**, and use the **External link** option to enter the web address (URL) to the website to link to, such as http://externaldomain.com.

Options for linking to files in Media

>> Place the cursor in the rich-text editor where you want the link, and drag the desired file (PDF, Word and so on) from a Media folder into the location. The filename becomes the link name (which you can edit, if needed).

Select the text where you want the link in the rich-text editor, click insert/edit link and use the Media option to select a media file to link to server.

Linking to files on a file server or an external website
Select the text where you want the link in the rich-text editor, click **insert/edit link** and use the **External link** option to add the path or URL to the file to link to. Fora file server, enter the complete path including the computer name, and the filename with file extension. The file storage network location must be properly accessible for this to work. Valid external prefixes are *http*, *https*, *ftp* and *file*.

Linking to an email adress
Select the text where you want the link in the rich-text editor, click **insert/edit link** and use the **email** option to enter the email address to link to. When you click the link, the default email client opens with the email address entered in the address field of the email.

Linking from an image
Select the image in the rich-text editor, click **insert/edit link** and add a link using any of the **Page**, **Media**, **External link** or **Email** options.

Anchor links
When you add an anchor, first create the anchor and then add the link to it from the link dialog box.

1. Select the text in the rich-text editor where you want to place the anchor.
2. Click Insert/edit anchor and enter a name for the anchor. Avoid using special characters or spaces.
3. Select the text in the rich-text editor where you want to link to the anchor.
4. Click insert/edit link and use the Anchor option to select the anchor to link to.

Link properties
The following examples show common implementations of link functionality on websites, with functionality similar to what is described for the link dialog above.

Image link
When you add images to content areas intended for images or blocks and so on, instead of adding the image in the rich-text editor, you can use an image link property which lets you select an image from a folder in Media, automatically placing it properly in the content area. See Adding and editing images.

Link collection

A link collection is a property where you can manage a group of links.

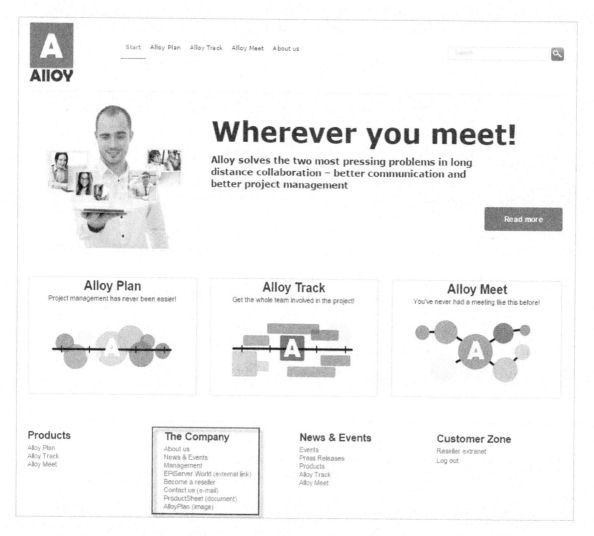

You can create links by dropping pages or media files in the link collection area, or by creating links through the link dialog. You can create links for pages, media files, external sources and email addresses. You can move links to change the order in which they appear.

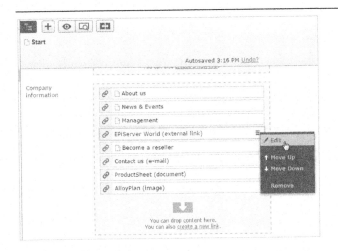

You also can edit the displayed name of the link.

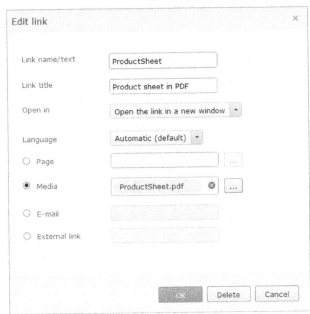

Commerce Linking to catalog entries

If you have Episerver Commerce installed, the link dialog box contains an option to select items from the product catalog when you create links. See Manage content in the Commerce user guide.

CMS Using forms (Legacy)

Web-based forms is a popular and frequently used feature on websites for creating questionnaires or registration forms for events. The built-in forms functionality in Episerver CMS is available through a form property, which is added to a page template or a block during implementation. On the Episerver sample site, the forms functionality is made available through a Form block.

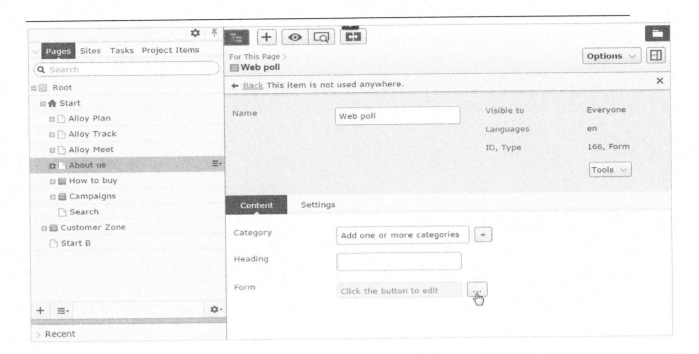

 You also can create block-based forms; see Episerver Forms in the online user guide.

Locate the **Forms** property (its location depends on yoursystem configuration), and click

The button to access the **Select Form** dialog, the "entrance" to forms management.

Managing forms

Forms are administered in the **Select Form** dialog box, from where you can view all web-based forms on the website, organize them in folders and use them in content. You can place forms in a selected folder when you edit or create them. You can delete form folders but you can only delete empty folders.

>> Click Edit to edit an existing form. You can save the existing form with a new name to keep the original.

>> Click Delete delete a form. You also can delete forms from within the Edit Form dialog box.

>> Click Select to select a form, and then Use to use the selected form in content.

>> Click No Form to remove a link to a form that is currently used in content.

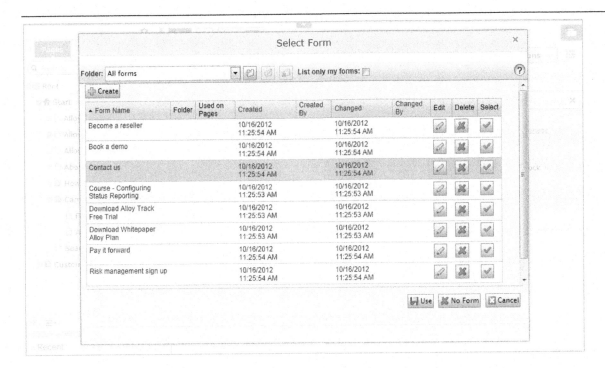

Creating a form

Start by creating the layout, then add the desired form fields. Complete the form by providing a name and defining the usage, then save it to become available for linking into content.

1. Creatingthelayout

Table Layout allows you to add rows and columns as desired to design the form. The table must contain at least one cell (row or column) before you can add any form fields.

>> Click Insert Row to insert a row above the row that is currently selected.

>> Click Add Row to add a row at the bottom of the table.

>> Click Delete Row to delete the selected row.

>> Click Insert Column to insert a column to the left of the selected column.

>> Click Add Column to add a column to the far right of the table.

>> Click Delete Column to delete the selected column.

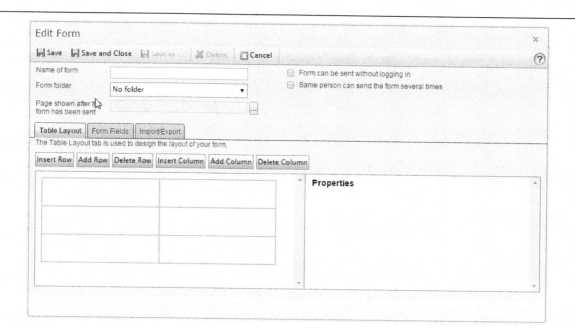

2. Adding form fields

After creating the basic layout, click **Form Fields** to add the fields. Click a cell in the layout, and select the desired type of field to add. You also can drag a desired property into the selected form table cell. You can add only one field in each table cell. Depending on the selected field type, properties are displayed for values to be entered. Save your changes when done with a of field properties.

If you chose to save the form data in the database as a posting result, the aggregated form data can be retrieved for viewing and exporting. Open the content (page or block) where you

have linked to the form in the All Properties editing view, and locate the **Forms** property. Select the **View data** option to access the form data.

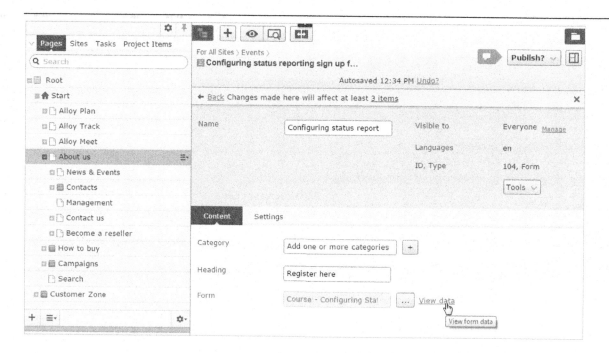

Forms data displayed is aggregated from all instances (page or block) where the form is used, if needed select a date interval to filter the data and click **Search**.

Before exporting, you can clean up the data by deleting any incorrect postings. Select the postings to export (click **Select all** to include all), and select the desired format to export to:

>> Export to Excel exports the data to a Microsoft Excel file. >>

Export to XML exports the data to an XML file.

Exporting forms

You can export forms between Episerver CMS websites. When you export a form, an XML file is created, which is then imported to the other website. Select a form to edit, go to the **Import/Export** tab, and click **Import** or **Export**.

CMS   **Inserting embedded media**

Adding embedded media to content works in the same way in both Episerver CMS and Episerver Commerce. Embedded media can be, such as a video or flash animations. Just as with images, the embedded media must be available on the **Media** tab to be able to link to the media file as described in Managing media. The most common file formats are supported on the Episerver CMS sample site. For information about other formats such as Quick time, Windows Media, and Real Media, see available accessibility coding standards.

 Depending on the type of media you select on the **General** tab, the attribute options vary on the **Advanced** tab. For information about Flash movie attributes, see available accessibility coding standards.

Insert embedded media as follows:

1. Place the cursor in the editor area where you want to insert your image.
2. Click Insert/edit embedded media image 🖼 on the editor toolbar.
3. In Type, select the type of media and associated format, such as Flash, Quick time or Windows Media. Flash is the default.
4. In File/URL, browse to select the media file in the File Manager.
5. In Dimensions, set the dimensions of the movie in pixels. Ensure that Constrain properties is selected to keep the proportions of the movie.
6. Click Insert and the media is linked into the page.

Adding dynamic content

Add dynamic content to a page by retrieving it from different **properties** fora page. The source of the dynamic content can be, for instance, text in the "main body"field (the editor area on a page), or the date when a page was saved.

For example, you can use dynamic content to display company facts and figures that are reused on multiple pages of a website. You also can combine dynamic content with a visitor group; see Personalizing content.

Dynamic content is not enabled by default. An administrator must enable it in the administration view. If it is not enabled, you will not see the **Dynamic content** button {} on the editor toolbar.

Add dynamic content from a **Page Property** as follows:

1. Open the page or block where you want to add the dynamic content and click the Dynamic content button {} on the editor toolbar.
2. In the Dynamic content window, select the plug-in that you want to use as a base for your dynamic content. An Episerver CMS standard installation includes the Page property plug-in, which is used in this example.
3. In Page to insert content from, select the page in the tree structure from which you want to display the data.
4. In Property to insert content from, select the property on the page from which you want to display data. In this example, you fetch data from the Main body property of a page.
5. Optional: In Personalization settings, you can click + to select the visitor group you want to have access to the dynamic content.

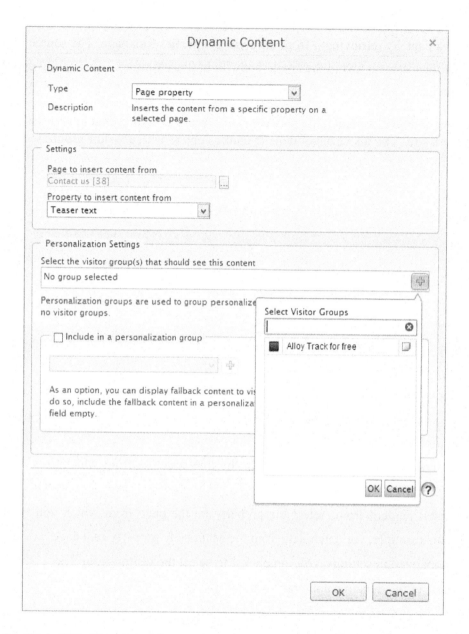

6. Click OK. The dynamic content appears as a box in the editor area. When this property is updated, all dynamic instances of the property in the content are automatically updated.

7.

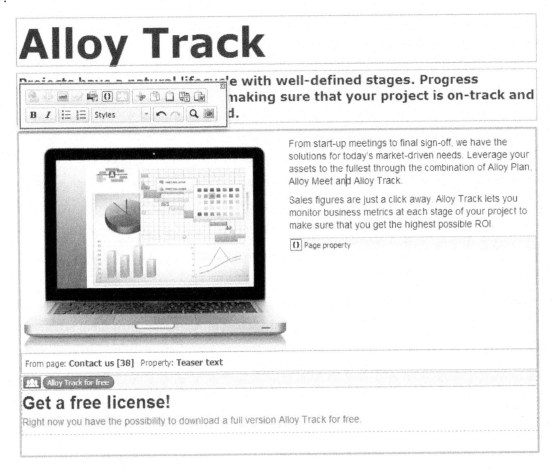

Preview the content and publish, or schedule for later publishing.

Editing dynamic content

You can cut, copy and paste dynamic content boxes in the editorarea, just as you can with any other object. Select the "box"with dynamic content in the editorarea, and click the **Dynamic content**toolbar button to edit. To delete, select the dynamic content box you want to delete and click **Delete**.

Cut and copy fora dynamic content box in the editorarea may work differently depending on the browseryou use. You may have to use eitherthe cut and copy editortoolbarbuttons, orthe right-click and cut and copy of yourbrowser, instead of the keyboard keys.

© Episerver

`CMS` *Previewing*

In Episerver, you can preview content while you are editing to view content as visitors will see it when published. The **preview** button [ ] hides the surrounding on-page editing frames and panes and you can navigate through yourwebsite.

The preview option can display different things depending on how you are working with Episerver CMS:

>> You are not using projects or you are using projects via the projects gadget. The preview option displays either the published version of each page, or if there is a newer draft, the primary draft version. Note that you may have a draft version in a project that is not set to primary draft, in which case the preview does not display the project version. Use the View settings > Projects option (see table below for details) to view your website as if the items included in a project were published.

>> You are using the projects feature. The preview option displays the active project. If no project is set as active, it displays either the published version of each page, or if there is a newer draft, the primary draft version.

View setting options	Button	Description
Visitor groups	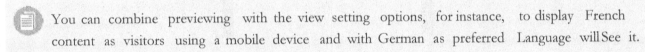	View the content as the selected visitor group will see it.
Media channels		Select a channel and/or a resolution to see the content as it will appear with the selected settings. Note that the options here are customized for your web-site.
Projects		Navigate and view the contents of a project to verify the display before publishing. **Note:** This option is not available if the projects feature is enabled. In that case, the preview button displays the active project. To preview another project, you need to change the active project.

You can combine previewing with the view setting options, for instance, to display French content as visitors using a mobile device and with German as preferred Language will See it.

EUG- Managing content |72

Comparing versions

In the compare view in Episerver you can compare content and properties between specific versions to see what has changed. When you compare content, the On-page edit view displays two page versions side-by-side. Properties are compared in the all properties edit view.

Toggle the **Compare different versions** button on the toolbar to turn the compare view on and off.

The **Select compare mode** button appears in the compare view and shows the selected compare mode; that is, if you are comparing content or properties. This selection is *sticky*, which means that whatever mode was used the last time you did a comparison is preselected the next time you turn on the compare feature.

Click this button to display a drop-down menu where you can change compare mode:

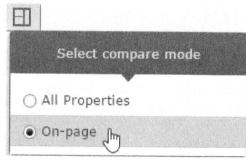

The notification bar displays two drop-down lists from which you can choose which versions to compare. By default, the draft is shown in the left pane and the currently published version in the right. All language versions of the content are listed by default.

Comparing content

When you compare content, the On-page edit view shows two versions side-by-side. You can scroll and resize the panes.

When you edit a published version in the left pane, a new draft is created and displayed in the version list. It works in a similar way as you edit content directly on the page, and when you are done, you can publish a draft, or republish a previous version.

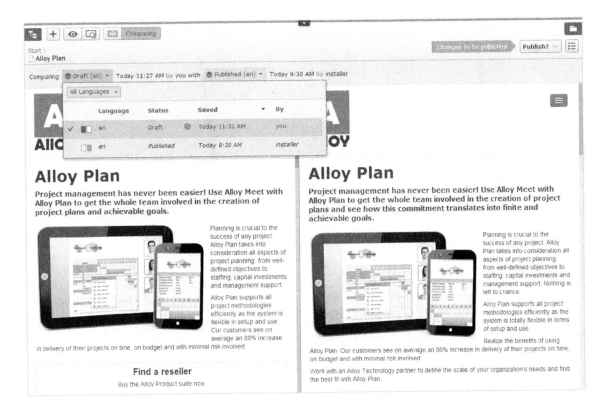

Comparing language versions when translating content

The current language is selected in the language selector list, and you can filter the versions by language. You can compare versions made in the same language, or in different languages.

By comparing versions made in different languages, you can translate the content in the left pane side by-side with the published version of the current language. You also can jump between languages to edit by selecting the current language in the version list of the left pane, and then switch language on the notification bar.

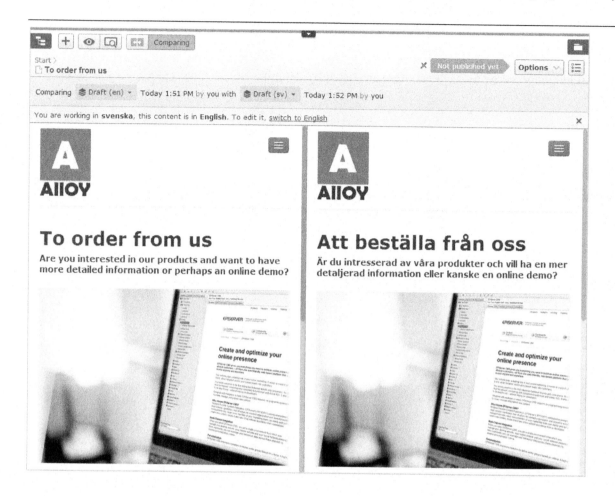

Comparing properties

When you select to compare all properties, the All Properties editing view displays the two compared versions of the properties side-by-side. They are displayed in the same tabs as they usually are, with the exception of the Basic info properties, which are displayed in a tab of their own instead of in the Basic info area. Tabs that contain changes between versions are highlighted with a yellow digit; the digit identifies the number of changed properties there are on the tab between the two compared versions.

In the compare properties view, the two property versions appear side-by-side. The latest version of the property is shown first, either to the left of or above the older version of the property. All properties that differ between the two versions are highlighted with a yellow background.

You can edit the latest version of the properties, and if you decide that you prefer the older version to the newer, you can click **Copy** and the older version is used in the newer version also. When you edit a published property, a new draft appears in the version list.

CMS *Publishing and managing versions*

Episerver has sophisticated support for advanced management of content creation and publishing involving multiple editors. The draft concept is central, ensuring that work-in-progress is never externally exposed until it is **actively published**. The publishing options you see depend on the content status and your access rights. Available actions, content status and notifications are indicated in the status bar at the top.

Publishing involves steps from creating a draft to publishing the final version, and managing versions. The steps apply to different types of content such as pages, blocks and media, or products if you have Episerver Commerce installed on your website.

 You can watch the following demonstration video, Publishing content. (4:18 minutes)

Publishing actions

When you create orupdate content, you can perform a numberof actions to create drafts, undo changes, set content ready forreview, publish directly orschedule publishing at a laterstage, and so on.

Creatingdraftsandautosaving

Whenever you create new content oredit existing content, a **draft** version is automatically created. This is not publicly available on the website until actively published. Changes to content properties are immediately **autosaved** by the system. Versioning is not used for drafts, which means that you and other editors can work on the same draft overtime but you still will have only one version of the draft.

You can manually create a draft from a previously published version orfrom a version that is scheduled for publishing at a later time. This is done from the **Options** menu.

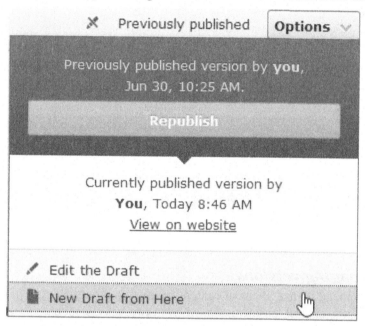

Undoing and reverting to published

While editing, clicking the **Undo** option in the page information area lets you undo changes to content that was previously autosaved.

» Select Undo to discard the changes done since the last auto save.

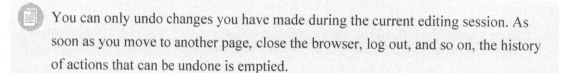
> You can only undo changes you have made during the current editing session. As soon as you move to another page, close the browser, log out, and so on, the history of actions that can be undone is emptied.

»

»

Select Redo if you discarded your changes through Undo and want to take them back again. Select Revert to Published to take back the latest published version, if the content was previously published.

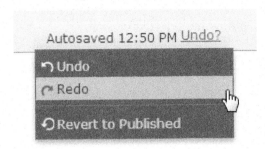

Previewing and Comparing
You can preview content appearance using the **Preview mode** option in the top toolbar. You can also preview content by language, visitor group or display channel if these are used on your website; see Previewing.

You also can compare different content versions by using the **Compare version** option in the top toolbar; see Comparing versions.

Publishing
When done editing, click **Publish?** at the top and then **Publish** (or **Publish Changes**, if you are editing previously published content). The content is immediately published and publicly available on the website, provided that no access restrictions apply. Click **View on website** to view the content as it appears on the website.

Setting Ready to Publish
If you do not have publishing access rights, or if you want your changes to be approved by someone else before publishing, use the **Ready to Publish** option to mark the content as ready for approval and publishing. **Withdraw and Edit** lets you take back content for further editing after sent for approval.

Removing scheduling and creatingnew drafts

Content that is scheduled forpublishing, is locked for editing. Select the **Remove Scheduling and Edit**option to interrupt the scheduled publishing and continue editing the selected version.

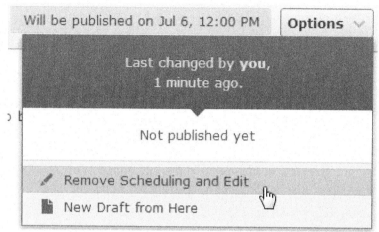

Selecting **New Draft from Here** creates a new draft, **based on the scheduled version**, which is still published at the scheduled time. You can continue working on the new draft, and apply publishing actions for this, as desired. An advanced scenario would be to apply multiple publishing occasions for different versions of a campaign page, having them replace each other in a desired order.

Publishing multiple content items

The projects feature and the projects gadget let you preview and publish multiple content items at the same time, such as a landing page, blocks and products (if you have Commerce installed)that are part of a campaign.

Managing versions

If you need to back track and use an older version of a page orif you are managing multiple language versions, there are a number of tasks you can perform from the version list by using the **More options** button found at the bottom of the version list. Click the column headers to sort the version list according to language, status and more.

 You need to add the versions gadget to the left or right panel to see the version list.

Viewing versions

Content can have the following status in the version list:

» Draft. Content that is a work-in-progress and is not yet subject to any publishing actions.

» Published. The most recently published version and the one publicly displayed. Only one published version can exist.

» Previously Published. One or more versions that were published before the latest published version.

» Ready to Publish. Content awaiting approval and publishing.

Rejected. A draft that was rejected by someone as part of an approval flow. You can edit and update the rejected content and then set to Ready to Publish again.

» Delayed Publish. Content scheduled to be published at a specified time. Expired.

» Content where a stop publish date and time is set and passed.

»

 You can define the number of stored content versions in the administration view. The default setting is 20.

Setting the primary draft

 Versioning works differently if you are working with projects. For a more detailed description on how Episerver handles versioning in projects, see Versioning when working in projects.

The **primary draft** is the draft presented in edit view, when accessing the content. Multiple drafts may exist, by default the latest saved edited version is the primary draft.

Use the **Set as Primary Draft** option in the version list to make another draft the primary one.

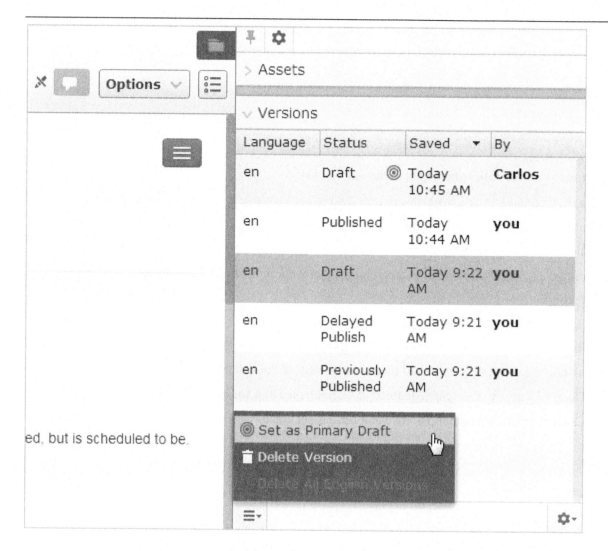

A content item that is not published can have only one draft. You can create multiple drafts from published versions, each draft is editable individually and you can schedule for publishing at different times. There are no versioning of drafts so there is always only one version of each draft.

Editing and deleting versions
The content version selected in the version list is loaded into the editing area, from where you can edit the content or perform other available publishing actions.

Select the **Delete Version** option to delete a version. Content versions are not supported by trash management. So, when you delete a version in the versions gadget, the version is permanently deleted.

 The version with status *Published* cannot be deleted, to do this another version needs to be published first. Deleting content versions cannot be undone. You can disable the ability to delete versions in the administration view.

Managing language versions

If the selected content exists in multiple languages, language versions are displayed with a language code. To filter versions fora desired language, select **Show Content in [language] Only** in the version list.

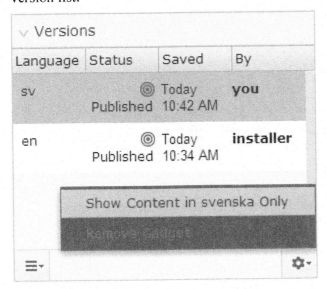

Republishing a version

To republish a previously published version, select the desired version and select **Republish** from the publishing options. When you republish content, for traceability reasons, a new version with a new times tamp is created, even if no changes were made.

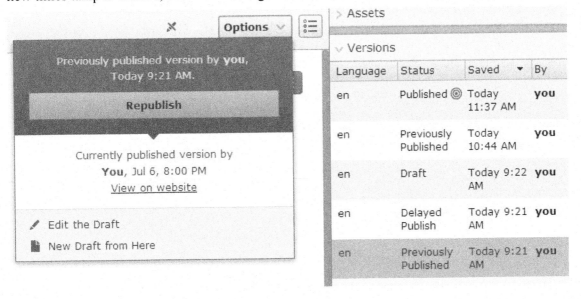

Working with shared content

As soon as you start to edit content, the content is marked as *currently being edited* notifying other editors to avoid version conflicts.

Mark as being edited
Even if content is marked as being edited, another editor can select the **Edit Anyway** option, and continue working with the draft.

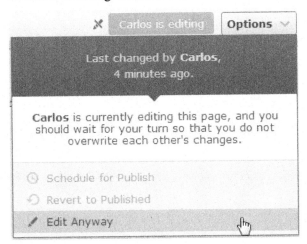

Permanently mark as being edited
The *currently being edited* markup setting is automatically cleared after some time. To keep this setting, you can set a *permanently being edited* markup through the All Properties editing view by selecting **Tools** > **Permanently Mark as Being Edited**. This setting remains until manually disabled (toggle the setting to disable).

Setting expiration of content

Normally, web content never expires, but you can set pages and blocks to expire at a certain time in the future or immediately. Expired content is not displayed on the website but remains in Episerver CMS. You can remove the expiration from the content to make it appear on the website again.

Setting an expiration time is done in the All Properties editing view by selecting **Tools** > **Manage Expiration and Archiving**. Select **Now** if you want expiration to apply immediately.

Archiving of expired content
Episerver has a built-in archiving feature where pages with a *set stop publish time* are automatically moved to a defined archive branch when the time has passed. For example, this is useful if you have news pages in a listing where you want to remove old news from the listing, but still keep the pages. Archiving expired content occurs in the **Manage Expiration and Archiving** dialog box by selecting the page branch to which you want to move the expired page.

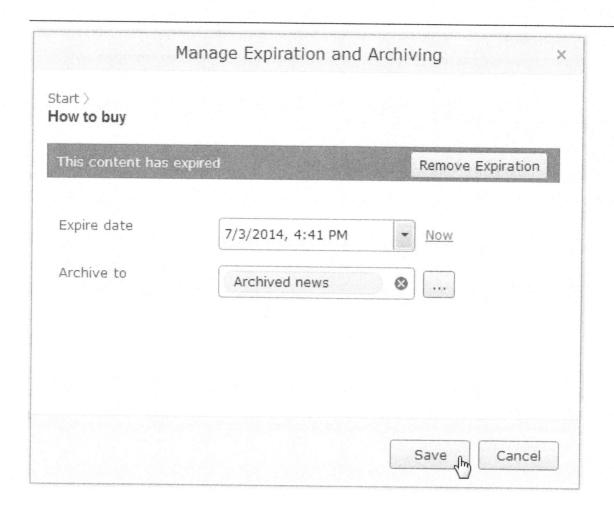

CMS   Managing content

Content can be pages and blocks in CMS, or product content from the catalog on an e-commerce site.

Content can also be assets such as images and videos, or documents in Word or PDF format. Episerver has a sophisticated version management features, allowing multiple editors to work with draft versions, before approving and publishing the content.

Content on a website can originate from different sources, depending on where on the site and by whom it was created.

>> Editors and marketers, or merchandisers can create content internally, on an e-commerce website.

>> A visitor community member can create content externally through interactive social features on the website, if these are available.

You can preview draft content before publishing, so that you can verify the content before publishing. When working with personalization, you can preview content the way it appears for different visitor groups. To further limit access to content that is work-in-progress, you also can set access rights for content from the edit view.

If you have content in multiple languages on your website, Episerver has advanced features for managing translation of content into additional languages, including the use of fallback and replacement languages.

Commerce Commerce-related content

See Managing e-commerce-related content in the Commerce User Guide, if you have Episerver Commerce installed.

Addons Optimizing content to improve search

See Working with content to optimize search in the Find User Guide, if you have Episerver Find installed.

CMS Projects

A project lets you manage the publishing process for multiple related content items. For example, you can add a landing page, blocks, pages and products (if you have Commerce installed)to a project and publish them, manually or scheduled, at the same time. Projects support management of content in different language versions, so you can manage translation of content items in projects.

In Episerver, there are two ways to work with projects.

>> Add the projects gadget to your user interface.

>> Have an administrator enable the projects feature for the entire site.

EUG- Managing content | 86

CMS *The projects gadget*

A project lets you manage the publishing process for multiple related content items, such as a landing page, blocks and products (if you have Commerce installed) that are parts of a campaign. Projects support content management in different language versions, so that you can manage translation of content items in the project view.

You can create new content or create draft versions of existing content, associate the content items with a project, and then either publish the project immediately or schedule it for later publishing.

Working with projects

Creating a project and adding content

Create a project from the gadget menu and add desired content items through drag-and-drop. You can prepare the draft versions of the content items first, and then create the project and add them, or the other way around. Use **Sort** in the context menu to sort content items fora better overview, and **Refresh**

Previewing project content

The preview option in the top menu has an option for projects where you can browse through included items, preview them as if they were published, and update them if needed.

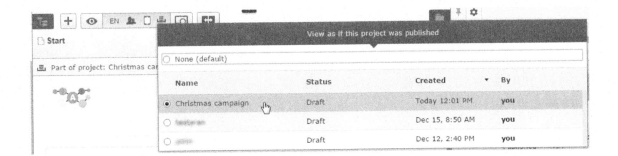

Publishing projects

To be able to publish a project, all included items must first be set to status **Ready to Publish**. You can do this for each item from the publishing menu when editing, or from the context menu in the **Project** gadget.

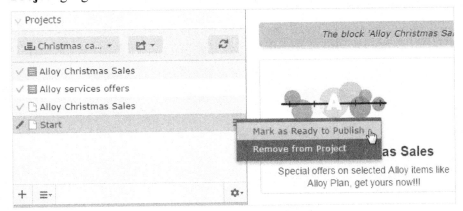

When all items are ready for publishing, you can publish the entire project directly, or schedule the project to be published later (if you have publishing access rights).

If you need to change content in a scheduled project, select **Remove Scheduling and Edit**, change the content and re-schedule the project publishing. Published projects cannot be edited.

Removing content and deleting projects

Remove a content item from a project by selecting the item in the projects gadget and then selecting **Remove from Project** from the item's context menu. Removing a content item from a project means that it is no longer associated with the project but it is not deleted from the website.

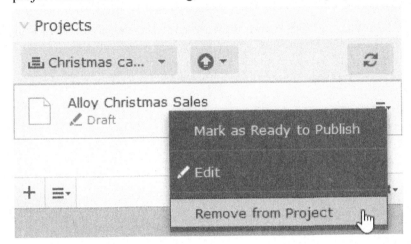

Delete a project by selecting the project in the projects gadget and then selecting **Delete Project** from the gadget's context menu. Projects are permanently deleted, but associated content items remain. When deleting a project scheduled for publishing, you have the option to either keep or remove the scheduling for each associated item.

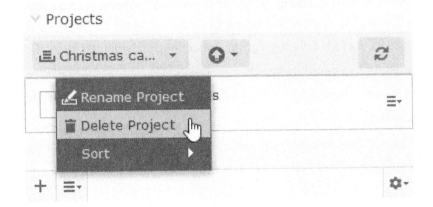

EXAMPLE: Publishing a campaign using Projects
This example creates a fashion sales campaign with multiple content items on a website with Episerver CMS and Commerce. The campaign will go live on a specified date, and contain a

landing page with a product listing block, two new products to be listed in the block, and a teaser block for the start page. You create the project first, and then add the content items.

1. In the Commerce catalog edit view, create a project for the campaign and name it Spring Collection.

2. Prepare draft versions of the catalog items in Commerce, create and edit the catalog entries and add product descriptions and assets as desired.

3. When done, drag the prepared catalog entries from the Catalog gadget to the Project gadget where you set the products to Ready to Publish before they are added to the project, but you can do this later.

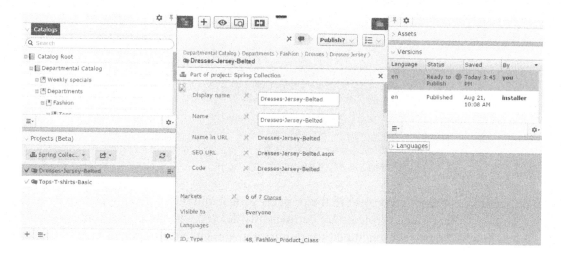

4. Switch to the CMS edit view and create the landing page for the Spring Collection. Add text and assets as needed, and drag the landing page to the Spring Collection project.

5. Create a block listing the products included in the spring collection, and include it in the landing page. Add the product listing block to the project.

6. Create a teaser block to be used on the landing page for promoting the new spring collection, and add the teaser block to the project.

7. Drag the teaser block to the start page, and add the start page to the project. The Spring Collection project now contains all the items to be included in the campaign.

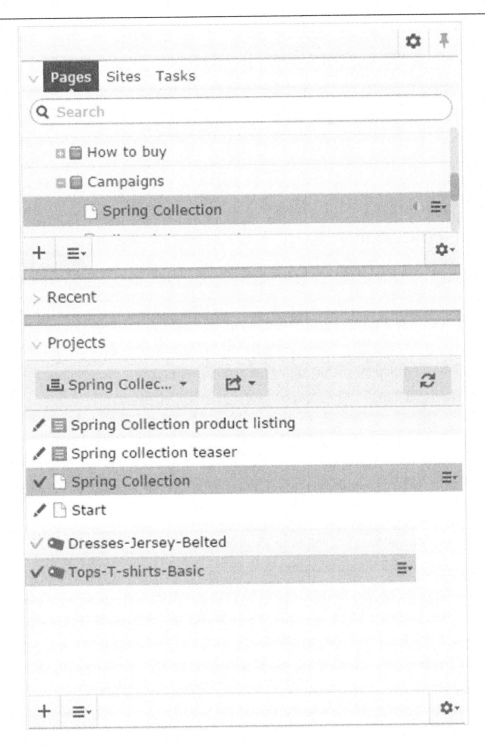

8. Preview the content items in the project, edit as needed and set to Ready to Publish when done.

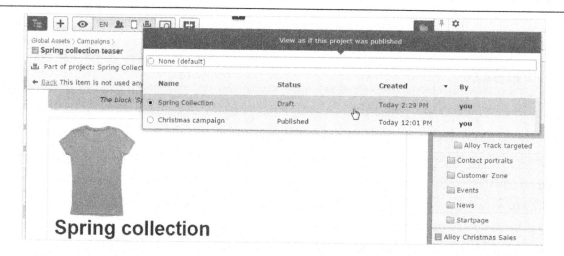

9. Schedule the project to be published on the defined go-live date for the campaign.

> You cannot edit versions of content items that are part of a scheduled project. For example, if you need to update the start page before the scheduled project is published, you need to create a new draft and then publish this. If you need to incorporate the same changes into the scheduled project version of the start page, remove the scheduling to edit.

EXAMPLE: Managing multiple content language versions using projects
This example creates a page with a registration form block for an event. The included items need to be available in English (the original website language), French and German, and you manage the translation using a project.

1. Create the page and the related forms block in English first.
2. Create a project and name it Spring Meeting.
3. Add the English versions of the page and the forms block to the project.
4. Enable and activate the desired languages (here French and German) on your website, if not already done.
5. Switch to the French language branch and create a French version for the page and the forms block.
6. Drag the French version of the content items into the Spring Meeting project.
7. Repeat the previous actions for the German language version. You now have six content items in the project; two for each language version.

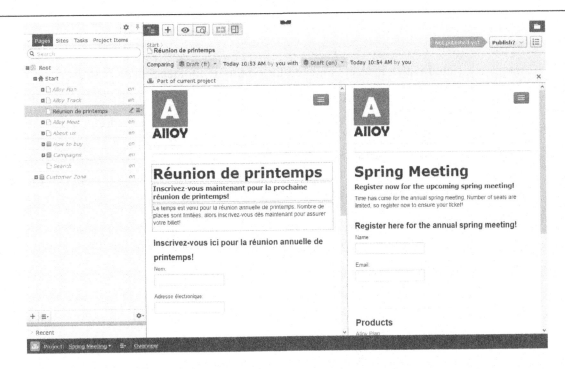

8. Translate the content items into French and German respectively.

   » Use compare to display the original English version when translating. »

   Use preview to verify the different language versions of the content.

9. Set all the content items to Ready to Publish when done, and either publish the project or schedule it for later publishing.

CMS   **The projects feature**

A project lets you manage the publishing process for multiple related content items. For example, you can add a landing page, blocks, pages and products (if you have Commerce installed) to a project and publish them, manually or scheduled, at the same time. The projects feature supports management of content in different language versions, so you can manage translation of content items in projects as well.

» The projects feature is enabled or disabled for the entire site and affects all users.

Editing actions, such as creating and updating items, automatically associate a content item with a currently active project. Exceptions to this rule are moving items in the structure, setting content to expired, changing access rights or language settings. These actions do not associate content items with the active project.

» Content associated with a project is locked for editing if another project is active.

» A version of content is associated with a specific project. This means that you can have a published version of an item not associated with any project; one draft of the same item belonging to a Summer campaign project, and another draft belonging to a VIP campaign project.

» You can add, remove and update existing items even if some or all of the items within a project are published.

» You can add, remove and update existing items even if some or all of the items within a project are published.

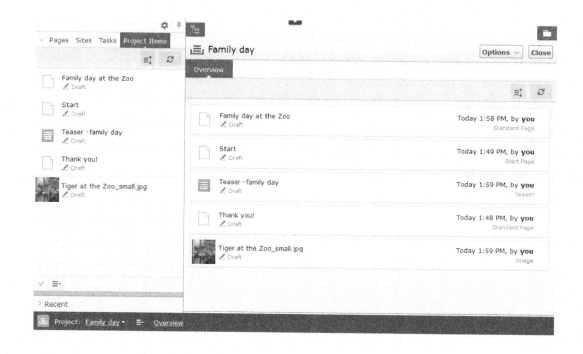

Projects user interface

When the projects feature is enabled by an administrator, a project bar is displayed at the bottom of the CMS window.

When you first access the edit view with the projects feature enabled, no project is selected in the project bar. When you select a project, it is preselected the next time you open the user interface.

If a project is active—that is, is selected in the project bar—all changes (creating a new page or block, updating existing content, uploading an image and so on) are automatically associated with that project.

If you select the option **None (use primary drafts)**, you can work with content items as usual without associating them with any projects.

From the context menu on the project bar, you can create, rename and delete projects.

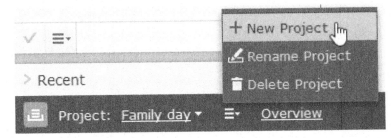

If you delete an active project, the project bar turns red. Associated items are not deleted but are no longer associated with a project.

The project overview

From the projects bar, you can open an overview that displays content items associated with the active project.

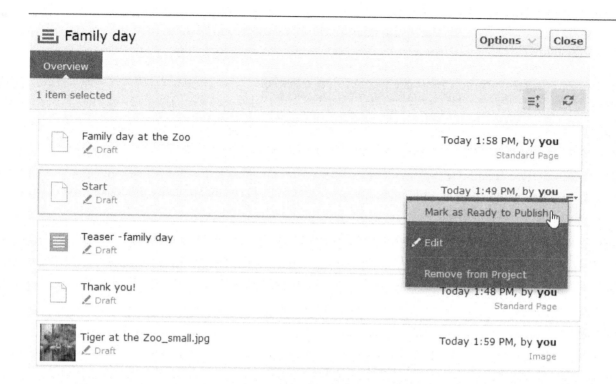

The overview shows details such as name, content status, content type, and time and date for latest change of each content item.

Each content item in the overview has a context menu from which you can set the item to *Ready to Publish*, open it for editing and remove it from the project, providing you have **Edit** access rights. The context menu button is displayed when you hover the mouse over an item.

You can select multiple items in the overview and remove them from the project orset them as *Ready to Publish* all at the same time. Common mouse and keyboard functionality for selecting multiple items is supported, except forCtrl + a which is not supported.

From **Options** in the overview, you can publish all items that are set to *Ready to publish* immediately or schedule them for publishing at a later time.

Show comments opens a view where you can select a project item and see a list of events connected to the item. You can add comments on each event and also reply to comments, see Working with comments for more information.

Use **Sort**≡↕ to order content items fora better overview, and **Refresh**↻ to reload the view if there are multiple editors working on the same project.

The project items navigation pane

The project items navigation pane provides quick access to items in the project. Double-click on an item to open it.

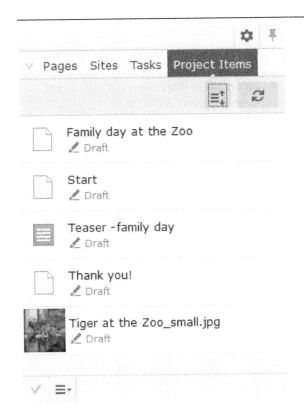

Each content item in the project items pane has a context menu displayed when you hover the mouse over an item; the menu options are the same as those in the project overview. You can select multiple items in the list the same way as in the project overview.

Versioning when working in projects

 Add the versions gadget to your user interface if you are working with multiple drafts and projects to see a list of the different versions.

It is only *one version of a page* that is associated with the project. This means you can have a published version of a page and several drafts, and any one of these versions can be associated with the project. If the associated version is the published version, the project overview displays **Published** for that item. If it is not the published version, the overview displays **Draft**, **Previously Published**, **Expired** and so on.

You can only have one published version, so if you publish another version of the page (that is, a version that is not associated with the project) *after* the version associated with the project, the project version is not published anymore and therefore set as **Previously published** in the project overview.

When you open an item and have a project active, Episerver CMS displays the version associated with the active project. If you open an item and do not have a project active, Episerver displays the version that is set as primary. You can see which version is the primary in the versions gadget; the **primary version** is marked with a target symbol 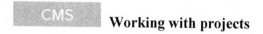. The primary draft is not necessarily the latest version. For information on primary drafts when not working with projects, see Setting the primary draft.

CMS **Working with projects**

Creating a project and adding content

You create a new project from the context menu on the project bar.

When you create the project, it is automatically set as the active project, which means that editing changes (such as adding new content items, updating existing content, uploading an image and so on) are automatically associated with the project.

Editing content in projects

You will see a notification if *the version you are working on* is associated with the active project.

Content not associated with the active project but with another project is locked for editing. However, even if that version is locked for editing, you still can create a new draft with the **New Draft from Here** button in the yellow toolbar. That draft is associated with the active project, or to no project at all if **None (use primary drafts)** is selected.

If you want to make a change that should not be associated with the active project, you have to select **None (use primary drafts)** or another project from the project bar. When **None (use primary drafts)** is selected in the project bar, you can create drafts, publish content and so on, as long as the content version is not associated with a project.

Content items that are part of a project remain so even after they are published.

Uploading media to projects

If you upload media while a project is active, it automatically associated with the active project.

 Media is never auto-published when you upload it to a project, not even if your system is configured for auto publishing of media (see the **Auto publish media on upload** setting in System settings in the CMS Administrator Guide). Media associated with a project need to be published like any other content item.

Previewing project content

The preview option in the top menu lets you browse through the website as if the project were published.

If you click **preview** , you preview the active project; to preview another project, change the project in the project bar. Previewing a project shows you either the draft associated with the project or, if there is no draft associated with the project, the published version.

Collaborating on projects

You can add comments on project items in the project overview and, for example, ask other editors to review an item. This feature is described in Working with comments.

Publishing Project items

From the project overview, you can publish multiple items that are set to status *Ready to Publish*. You can set items to *ready to Publish* either from the publishing menu while editing an item, from the context menu in project overview, or from the Project Items navigation pane. You can select multiple items and set them to *Ready to Publish*, both from the overview and the navigation pane. Note that you publish the content items associated with the project, not the project itself.

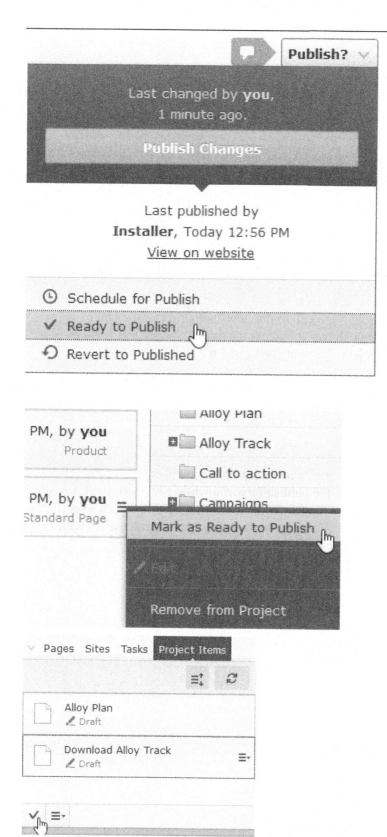

THE MISSING MANUAL

When items are ready for publishing, you can publish them directly from **Options**, if you have publishing access rights, or schedule the project to be published later.

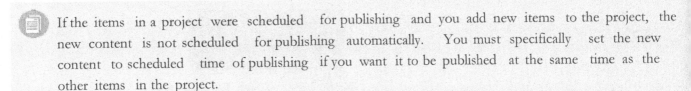
If the items in a project were scheduled for publishing and you add new items to the project, the new content is not scheduled for publishing automatically. You must specifically set the new content to scheduled time of publishing if you want it to be published at the same time as the other items in the project.

If you need to edit content that is scheduled, select **Remove Scheduling and Edit**, edit the content and reschedule.

You can continue working with a project after it is published.

Removing content and deleting projects

Removing content from projects

Remove a content item from a project by selecting the item either in the project overview or from the project items navigation pane and then selecting **Remove from Project** from the item's context menu. Removing a content item from a project means that it is no longer associated with the project but it is not deleted from the website.

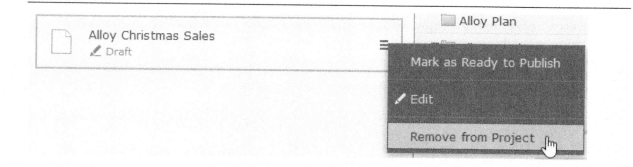

Deleting projects

Delete a project by selecting the project in the project bar and then select **Delete Project** from the context menu. Associated content items are not deleted but are no longer associated with a project. You cannot recover deleted projects. When deleting a project with items scheduled for publishing, you can keep or remove the scheduling for each item.

CMS **Examples for using the project functionality**

EXAMPLE: publishing a customer event using projects
This example creates an invitation to a customer event, including a registration form (using a block), a thank you for the registration-page, and a teaser block for the start page. All content items for the event are scheduled to publish at the same time.

1. Create a new project for the event from the project bar and call it Customer event.
2. Prepare a page with information on the event. The page is automatically added to the project. You can set the page to Ready to publish or do that at a later stage.

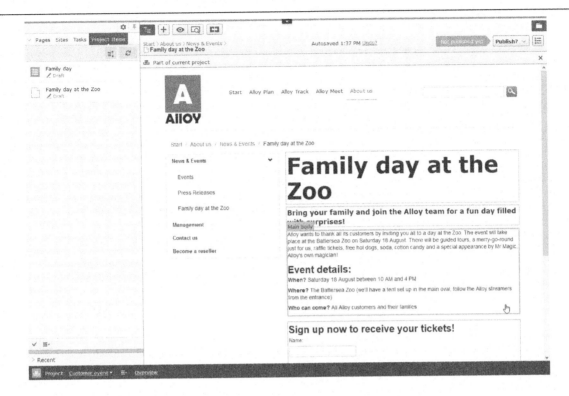

3. Prepare a thank you-page.

4. Create a registration form. On the sample site, this is done with a form block. Set up the form so that a visitor who registers for the event is directed to the thank you-page.

5. Create a teaser block to use on the start page for promoting the customer event and drag it to the start page. The Customer event project now contains all items related to the event.

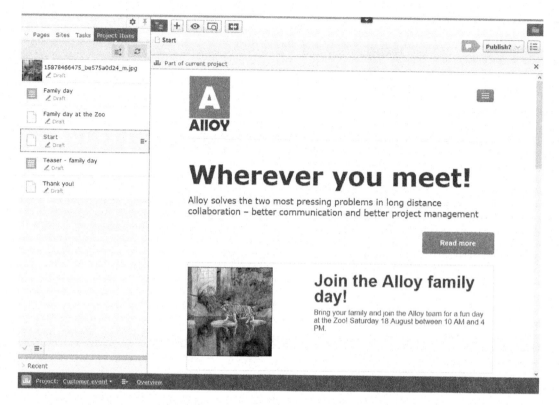

6. Preview the pages by clicking preview 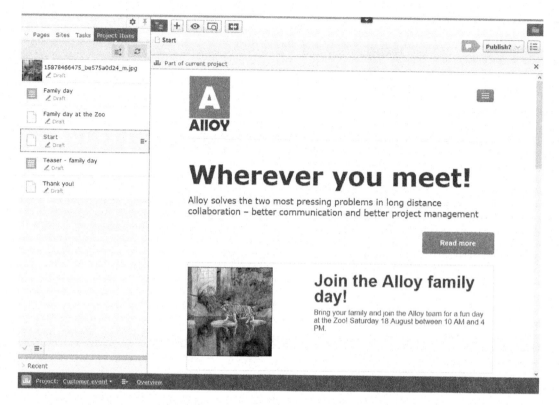 .

    » Does the start page contain the teaser?

    » Does it lead to the customer event detail page?

    » Fill in the form and make sure that you are directed to the thank you-page.

7. Go to the overview and select all items by pressing Shift on your keyboard and

    Selecting the top and the bottom items.

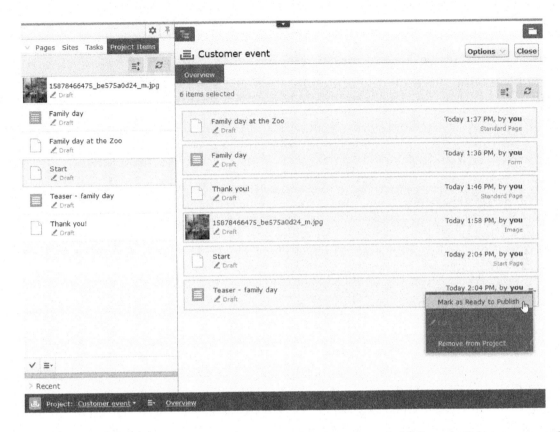

8. From the context menu of one the items, choose Ready to publish and all items are ready to be published.

9. Select Options > Schedule Items for Publish in the overview.

10. Select 1Augustat11AMand click Select.

EXAMPLE: managing multiple content language versions using projects
This example creates a page with a registration form block for an event. The included items need to be available in English (the original website language), French and German, and you will manage the translation using a project.

1. Create a project and name it Spring Meeting.
2. Create the page and the related forms block in English first. They are automatically associated with the project.

3. Enable and activate the desired languages (here French and German) on your website, if not already done.
4. Switch to the French language branch and create a French version for the page and the forms block.
5. Repeat the previous actions for the German version. You now have six content items in the project, two for each language. You can see all six items in the Project Items

    Navigationpane.

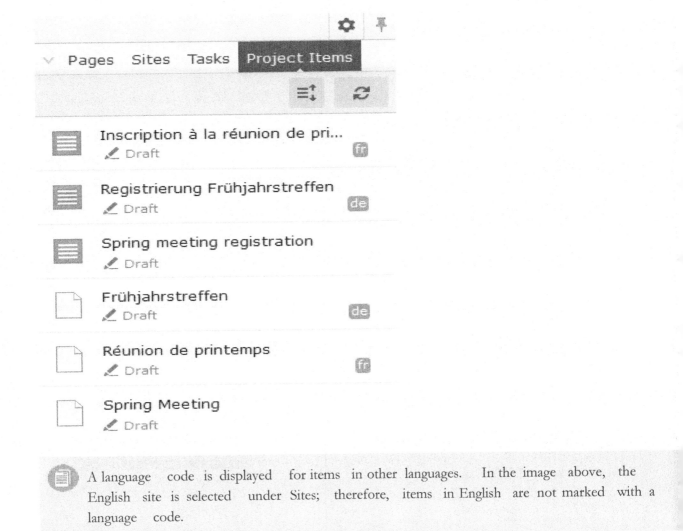

A language code is displayed for items in other languages. In the image above, the English site is selected under Sites; therefore, items in English are not marked with a language code.

6. Translate the content items into French and German respectively. Use compare to display the original English version when translating. Use preview to verify the Different language versions of the content.

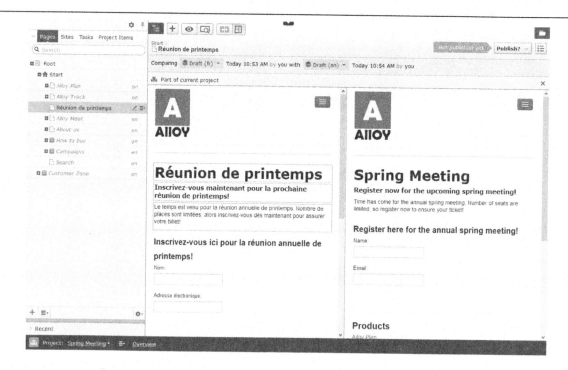

7. Set all content items to Ready to Publish when done, and either publish the project or schedule it for later publishing from Options in the project overview.

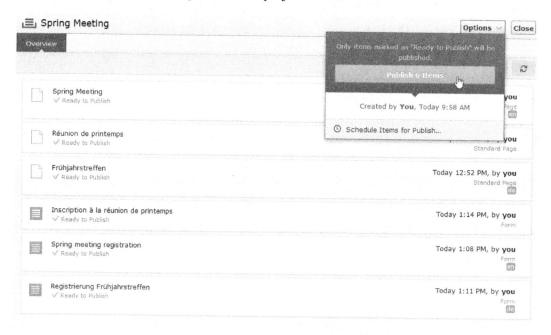

THE MISSING MANUAL

CMS *Working with comments*

Comments facilitate collaboration between editors of a project. You can add comments on a project or on specific project items or actions, add information on changes you have made, other editors to review the item and so on. Other editors can view your comments and reply to them, and also add their own. You can also tag users in a comment.

> A comment is only connected to the current version. As soon as a new version of the project item is created, the comment connected to the previous version disappears.

If another user has tagged you, made a comment on one of your actions or replied to one of your comments, you are notified in the user interface. The bell icon in the toolbar displays the number of new notifications you have. Click the icon to display a list of notifications. From the notification list, you can go to the project overview to read the comment.

 If you go to the project overview to read a comment, the project is automatically activated. If you do not want to continue working in the project, you have to deactivate it.

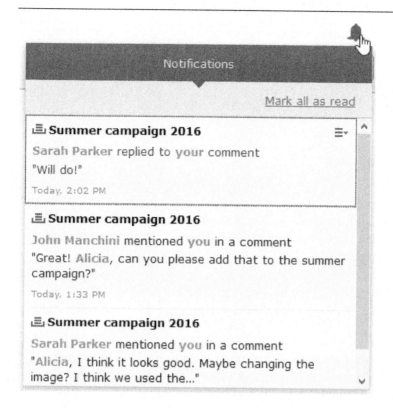

If your system is configured to use email notifications, you receive an email when someone else replies to one of your comments or replies, or if you have set a project item to **Ready to Publish** and someone else comments on this action. You also are notified if someone tags you in a comment. How often these notifications are sent depends on the system configuration.

Adding a comment to a project or project item

1. To add a comment to:

    >> a project, select the Project Comments tab in the project overview

    >> a project item, select Show comments in the Items tab of the project overview and then a project item. Items that already have comments have a comment icon

2. Add a comment in the comment text box.

3. To tag another user, enter @ followed by the user name. Select the user from the displayed list of suggested users. The tagged user receives an email notification and a

4. notification in the user interface when logging in. You can tag any of the available users but the tagged user still needs access rights to the project item to see the actual comment.

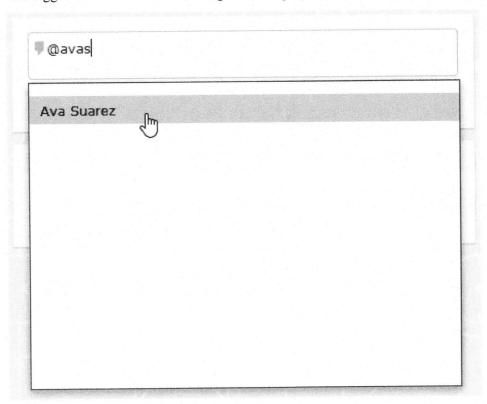

5.Press Enter or click post

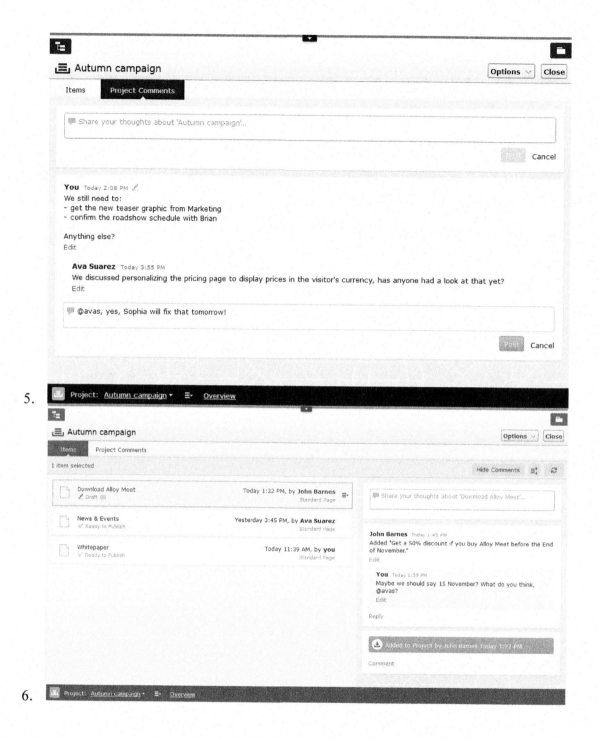

5.

6.

7. **Replying to a comment**

8. Click **Reply** on the comment you want to reply to and enter your comment in the text box. Press **Enter** or click **Post**.

Editing a comment or reply

You can edit comments and replies that you have created yourself. Click **Edit** on the comment or reply you want to change. Press **Enter** or click **Save**. Edited comments or replies are marked with a pen symbol ✎.

Deleting a comment or reply

You cannot delete comments or replies yet.

CMS   Structuring the website

In the Episerver CMS, the page tree structure is located in edit view under **Pages**. At the top of the structure is the **root** page, usually with one or more start pages directly underneath. The structure of the website is made up of pages. By default, the page structure is reflected in the navigation menus. To simplify navigation, you should limit the submenu structure to a maximum of three levels.

The page tree

By moving the mouse pointer over a page in the tree structure, information about the page, such as ID and page type, is displayed. A set of page tree symbols provides additional information about the structure.

Moving, copying and removing pages

Moving a page
Use drag-and-drop to move a page, or select **Cut** in the context menu for the page you want to move, and select **Paste** for the destination page. You can also move pages by using keyboard commands Ctrl+x or Cmd+x, and Ctrl+v or Cmd+v.

 When you move a page, internal links are redirected to the new location and are not broken. However, external links pointing to the moved page will be broken.

Copying a page
Select **Copy** in the context menu for the page you want to copy, and select **Paste** for the destination page. You can also copy pages by using keyboard commands Ctrl+c or Cmd+c, and Ctrl+v or Cmd+v.

Subpages and associated media files in local page folders are copied also, and the links point to the new copy of the page. Settings, such as dynamic properties and categories, are copied also with the new page.

 When you copy and paste a page under the same node, the **Name in URL** property of the copied page is typically named [Name in URL1], which you need to change after copying.

Removing content

Removed folders, pages, blocks and media files are moved to trash, and are not publicly available on the website. Remember to update any links to removed content. When you remove a page, all underlying pages are removed also. See Deleting and restoring content.

Sorting page order in menus

The pages in the tree structure are sorted according to a predefined sort order. By default, the page that was created most recently is placed at the top of the tree structure, for example, in news listings. There also are other options for sorting, such as alphabetically or by sort index. The last option lets you control the sorting through an index defined on each page.

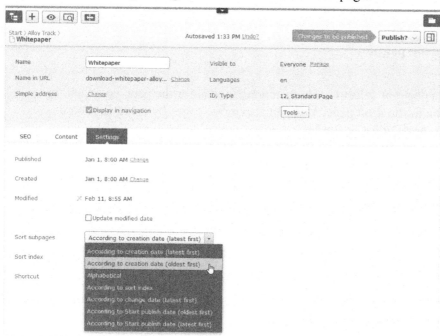

The sort order is set for the parent page of a branch, and is inherited by the subpages:

Select the parent page of the branch in the structure where you want to set the sorting.

1. Edit the page and select the Settings tab.
2. Select sorting criteria under Sort subpages.
3. Publish the page for the changes to take place.

Sorting according to sort index

If you want to control exactly how the pages in the structure are sorted, select the **According to sort index** option for the parent page. Then give each child page a unique sort index number, and they are sorted in ascending order according to their number, with the lowest number on top.

Change the sort order of pages by dragging the page and dropping it where you want it in the tree structure. Sorting pages with drag-and-drop is only of interest for branches that are sorted with sort index.

>> If you move a page into a branch that is not sorted according to sort index, you get prompted to move the page and at the same time apply sort index as sort order for that branch. Confirm with

OK. When you drag a page into a new position in a page tree branch, the sort index is automatically recalculated to fit the sort order of that branch.

>> If you drop a page under a different parent page (with sort index set as sort order), the page is first moved or copied, and then sorted. The page remains in the tree in the position where it was first dropped.

 The pages that you move are saved again, meaning that you must have publishing rights to use drag-and-drop for sorting. You also must have publishing rights for the page branch with sort index to which you move a page.

Sorting according to sort index manually

You can set the sort index manually for each child page. Open the page for editing, select the **Settings** tab and change the number in the **Sort index** box. The sort index number must be an integer, but there are no other restrictions. You should work with whole tens or hundreds to insert additional pages in between existing ones in the structure. Publish the page to apply all changes to the structure.

Deleting and restoring content

Episerver has advanced support for restoring deleted content such as pages, blocks, folders and media files. When you remove content, you move it to trash; you do not delete it. From there, you can either restore the content or permanently delete it.

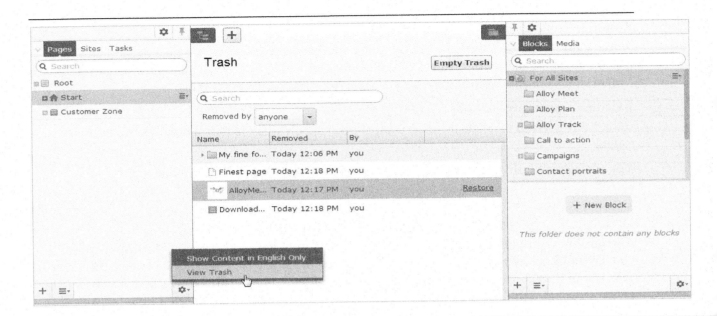

Episerver Commerce does not support trash management when deleting catalog entries.

Moving content to trash

Select the content to delete, and select **Move to Trash** from the context menu.

Content that is moved to trash is automatically unpublished from the website. When moving content to trash, you receive a notification if the content is linked to from other content on the website, because the deletion might result in broken links.

Restoring content

Select **View Trash** from the pane settings in the navigation or assets pane. Select the desired content in the list and click **Restore**. The content is restored to its original place and republished. You must restore content to edit it.

Deleting content permanently

Click **Empty Trash** to delete the trash content permanently (may require administrative access rights).

You can automatically empty the trash at a regulartime interval using a scheduled job.

CMS Setting access rights from edit view

Administrators generally manage website access rights from the administration view. However, if you have **administer** rights, you can set access rights fora single page ora block from the edit view. This is useful when you need to publish an item to verify the final result, but you do not want it to be publicly visible. Setting access rights from the edit view only affects the **selected item** (page or block).

To set access rights, open the item in the All properties edit view. The **Visible to** option displays **Everyone** for content that is publicly available on the website, and **Restricted** if access limitations apply.

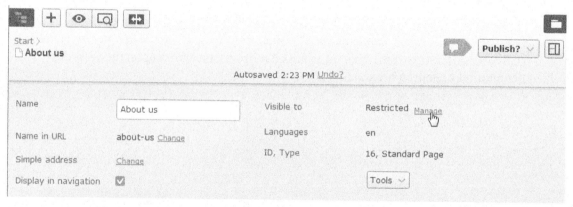

Click **Manage** to change the settings in the Access Rights dialog:

If access rights are inherited from the parent page, clear **Inherit access rights from parent item**, and click **Add Users/Groups** to define new access rights. Add access rights as desired and save the settings.

For example, removing read access for **Everyone**, as in the example above, hides the published page from the public, but it is fully visible and editable for the **SiteEditors** group (and **Administrators**).

 You must belong to a group with **Administer** access rights to define access rights from the edit view. This setting does not provide access to any other administration options in Episerver CMS.

See Access rights in the CMS Administrator User Guide for information about working with access rights in Episerver CMS.

`CMS` **Assets**

Assets can be, for instance, content of the type media files, images, documents, blocks or products from the catalogs in Episerver Commerce. Assets are available from the assets pane in both CMS and Commerce, making it easy to drag-and-drop items, such as images, blocks or products into, for instance, a CMS page.

You can work directly with content from the assets pane, for instance, edit images or blocks, or create folders to organize content items. The context menu will provide different options depending on the type of assets selected. How to work with content items in the assets pane is described in the sections Folders, Media and Blocks.

 By default, the assets pane in a standard Episerver installation will contain **Blocks** and **Media** with the addition of **Catalog entries** for Episerver Commerce. Since the assets pane is a plugin area, there might be other asset types available in your installation.

`CMS`

Folders in the assets pane in Episerver are used for organizing content, such as media files (for instance, images, videos and documents), as well as blocks. You can have folders with content that can be shared between all websites in a multi-site scenario, or you can have folders with content that will only be available fora specific website, or a page or block.

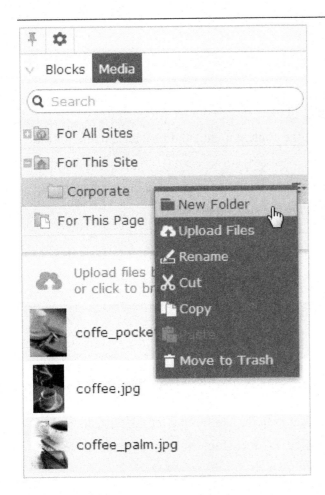

Note that by default, **media and blocks share the same folder structure**, meaning that if you create a folder under Media, the same folder is also created under Blocks.

Depending on your implementation, you may have the following predefined folders:

>> For All Sites. Content here is available for all websites in a multi-site installation.

>> For This Site. Content here is available only for a specific website in a multi-site installation. >> For This Page or For This Block. Local folder where content is available only for a specific page or block, and cannot be accessed from other pages or blocks. Useful, for instance, if you have images for a specific purpose which must not be used elsewhere.

 If you have saved an image in the local folder and then copy the page content, including the image, and paste it into another page, you may still see the image in the page. However, this is due to browser caching, the image is not copied to the new page's local folder and the link is in reality broken.

Local folders are not available forcatalog content in EpiserverCommerce.

Creating, renaming and moving folders

>> To create a new folder, select the desired folder in the structure under which you want to add a new folder. Select New Folder in the context menu, and provide a name for the folder.

>> Select Rename in the context menu for the folder you want to rename, and enter a new name.

>> Use drag-and-drop or Cut/Paste to move folders in the structure.

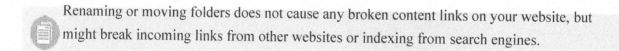
Renaming or moving folders does not cause any broken content links on your website, but might break incoming links from other websites or indexing from search engines.

Deleting folders

Select the folder you want to delete, and select **Move to Trash** in the context menu. The folder with its content will be moved to Trash, from where it can be restored. Local folders cannot be deleted.

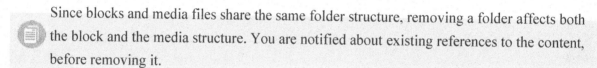
Since blocks and media files share the same folder structure, removing a folder affects both the block and the media structure. You are notified about existing references to the content, before removing it.

Setting access rights for folders

The predefined global folder is available to everyone by default. Local folders inherit the access rights from the content (page or block) to which they are associated. It is possible to define access rights for specific folders in a structure. Setting access rights for folders is done from the admin view in Episerver, in the same way as for pages in the page tree structure.

Managing folders in multiple languages

Folders are not language specific, and the folder structure for blocks and media will look the same regardless of the language selected under the **Sites** tab in the navigation pane. This means that you cannot create language versions for folders, but you can, for instance, use a language code when naming them.

CMS   Media

Media in Episerver are files that can be, for instance, an image, a document (such as a pdf document or a Word document), a video ormp3 files. Media is managed from the media library on the **Media** tab in the assets pane. Here you can create folders and upload media files. You can then make use of your media by dragging them into an Episerver CMS page or a block, or associating them with a product in Episerver Commerce.

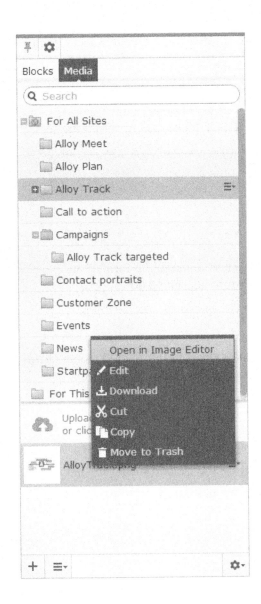

Searching for media

Use the search field at the top of the pane to enter search criteria and retrieve media files. Clicking a search result expands the folder where the file is located. To browse for media files, click a folder to expand the folders and content beneath it.

Uploading media

Media files are most easily uploaded through drag-and-drop from a file location on your computer to the upload area. You can also click directly in the upload area to add files. Or, you can select **Upload Files** in the context menu for the target folder to which you want to upload files.

 Depending on your implementation, media files may not be automatically published when uploaded. To implement automatic publishing of uploaded media, editors who upload must have publish access rights for the folder to which the media is uploaded. See Access rights for more information.

Previewing media

Media files in list views are represented by thumbnail images. Common image file formats are rendered for preview by default in Episerver, but other rendering formats can be developed.

Downloading media

Select the desired media file in the **Media** structure, and select **Download** in the context menu. Or, if you are previewing the media file, select **Download this file** from the **Options** combo button.

Editing metadata for media

Available metadata fields depend on the implementation; for images they can be, for instance, photographer, description and copyright information. Select **Edit** for the desired media file in the **Media** structure, and then the All Properties editing view to edit the metadata properties.

Renaming media

Select the media file in the **Media** structure and then the All Properties editing view, and change the **Name** and the **Name in URL**.

 Renaming a folder or media file changes its URL. This does not break internal links on the website, but incoming links from external websites may break.

Replacing media

To replace an existing media file with another, upload a new file with the exact same name to the same folder as the file you want to replace. A replaced media file is published immediately, affecting all places on the website where the file is used.

Managing media file versions

Versions for media files are managed in the same way as for other types of content, that is, by using the **Versions** gadget. Refer to Publishing and managing versions for more information.

CMS Blocks

Blocks are pieces of content that can be reused and shared between websites, while being maintained in one place only. Typical types of content blocks are campaign teasers and banners, videos, news feeds and contact forms. Just like for pages, you can have different block types, for instance, an editorial block, or a form or page listing block.

Blocks are managed from the **Blocks** tab under the assets pane in Episerver CMS, where you can create new blocks and organize them in folders. You can then utilize blocks by dragging them into the content area of Episerver CMS pages. You can manage block versions like other types of content, and blocks can also be personalized to be displayed for selected visitor groups.

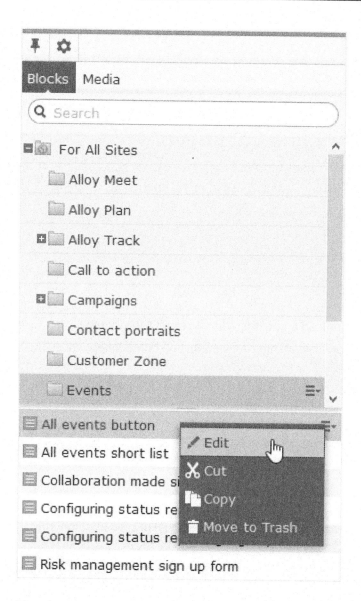

Use the search field at the top of the pane to enter search criteria and retrieve blocks. Clicking a search result will expand the folder where the block is located. To browse for blocks, click a folder to expand the folders and content beneath it.

Creating a block
Creating a block from the Blocks tab in the assets pane
When using this option, the block is saved in the block folder structure, and it is available for other pages on the website.

1. Select the folder in the structure under which you want create a block, and select New Block in the context menu, or click the Add button.

2. Select the block type among those that are available, and provide a name for the block.

3. Depending on the type of block, add content as appropriate.

4. Publish the block immediately or schedule for publishing later. Unpublished blocks are not visible to visitors, and appear dimmed out in edit view when added to a content area.

Spring is here!

Take advantage of great spring offerings.

Creating a block directly from a content area
When using this option, the block is saved in the **For this page** folder for the selected page, which means that it is not available on any other pages on the website.

1. Click Create a new block in the content area, either from the On-page or the All properties editing view.
2. Select the block type among those that are available, and then name the block.
3. Depending on the type of block, add content as appropriate.
4. Publish the block immediately or schedule for later publishing.

When creating a block, clicking **Back** takes you back to the page or block you were previously working on.

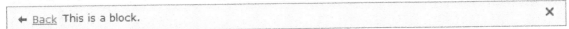

Editing a block

You can edit blocks either directly from the **content area** where it is being used, or from the **Blocks** tab in the assets pane.

1. Select the desired block to edit, and select Edit in the context menu.

2. Depending on the type of block, change the content as appropriate.

    If you want to rename the block, use the All Properties editing view.

3. Publish the block immediately or schedule for the changes to be published later.

Using blocks in content

Blocks can only be added to content areas that support blocks. In edit view, select the desired block in the assets pane, and drag it into a content area of a page. A green frame indicates where it is possible to add blocks on the page.

You can add several blocks to the same area. Drag the block above or beneath an existing block, and drop it when the separator appears. The blocks can be rearranged later. It is also possible to add blocks to a content area from the All Properties editing view.

```
                    Large content area

    ┌─────────────────────────────────────────────┐
    │ 📄 Alloy Meet jumbotron                       │
    └─────────────────────────────────────────────┘

    ▶ 👥 Personalized Group

    ┌─────────────────────────────────────────────┐
    │ 📄 Alloy Meet teaser                          │
    └─────────────────────────────────────────────┘
    ┌─────────────────────────────────────────────┐
    │ 📄 Alloy Plan teaser                          │
    └─────────────────────────────────────────────┘
    ┌─────────────────────────────────────────────┐
    │ 📄 Alloy Track teaser                         │
    └─────────────────────────────────────────────┘

                        ⌄

    You can drop pages and blocks here, or create a new
                        block

                                         [ Done ]
```

Like blocks, **pages** from the page tree can also be dropped into a content area. Depending on how the page template is built, the content of the selected page is rendered in the content area.

Blocks can also be added to a **rich-text editor area** through drag and drop.

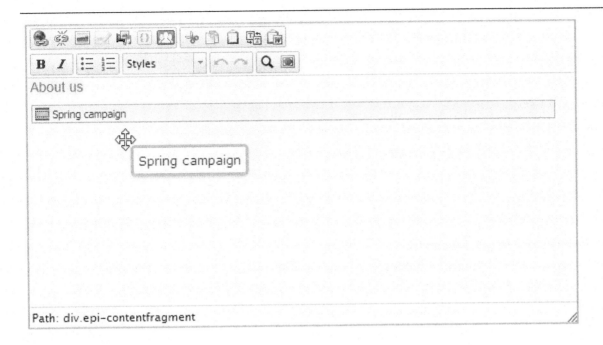

To **remove** a block from a content area, select **Remove** from the context menu.

 You can **personalize** blocks to display targeted information to selected visitor groups, see Personalizing content. Personalized blocks are not displayed in the edit view. Select a content area to display personalized blocks.

 You cannot link to blocks since they do not have a web address (URL). However, you can create links to otherpages and media files if the block contains the rich-text editor (XHTML string property).

Arranging blocks in a content area

You can change the display orderof blocks by **rearranging** them in the content area, eitherthrough drag-and-drop, orby selecting **Move up** or **Move down** in the context menu.

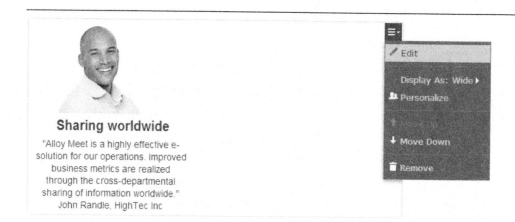

Displaying blocks in different styles

You can select display options for blocks on a page in different sizes and styles. The rendering of blocks needs to have built-in support for managing different widths, in order for the content to be properly displayed.

The following options are available:

» Automatic. Select this option to display the block using an appropriate built-in style option selected by the system.

» Manually. Select this option to display the block using the specific style option, for instance, presets such as Full, Wide, or Small, for the specific context where the block is used.

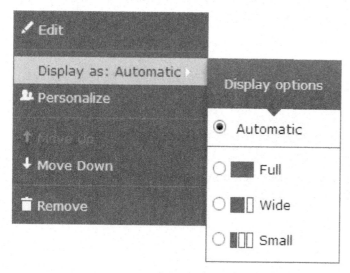

Moving, copying and removing blocks in folders

Moving, copying and removing a block works in a similar way as for pages by using the context menu. Since blocks and media files share the same folders, removing a folder from the tree structure affects all content within the folder. If any block or media within a folder is used on the website, you are notified about the usage before the content is moved to trash.

A block is no longer available on the website once it has been moved to trash. You can see removed blocks by selecting **View Trash** from the context menu of the block gadget.

Versions, content languages and access rights for blocks

>> Versions for blocks are managed in the same way as for other types of content. When you update the properties for a block, a new version will be created, which will be listed in the versions gadget. Refer to Publishing and managing versions for more information.

>> Content languages for blocks are managed in the same way as for other types of content, refer to Translating content for more information.

>> Access rights can be defined for creating and viewing blocks. This is done directly for a specific block in the All Properties editing view, or for an entire block structure from the admin view. From code it is also possible to restrict the block types that can be added to a content area. Refer to Setting access rights from edit view in this user guide and Access rights in the CMS Administrator User Guide for more information.

CMS   Managing multiple languages

Many large websites display content in several languages. Episerver has powerful support for Multilanguage management, including the possibility to translate content into a wide range of languages, defining fallback languages for non-translated content, as well as switching language for the editorial **user interface**.

How does Episerver know which language to display to visitors? Episerver enforces the language to be visible in the URL, either in the path or the domain part of the URL. When a website visitor selects a language option (if available), content in that language is displayed. Alternatively, the preferred content display language may be detected by the browser used by the visitor. If content does not exist in a selected language, a fallback procedure may be applied.

Enabling content languages

Usually a website has a default or "master" content language set up at the time of installation. In addition to this, you may add multiple content languages as required for your website. The enabling of languages is done in the steps described below. **Note** that you need administrative access right to access the administration interface in Episerver CMS.

1. Enabling a language on the website

This step activates the language to make it available for further configuration in CMS and Commerce.

1.   In the admin view, go to the Config tab > Manage Website Languages.

2.   Click on the desired language in the list (you can add a language if the desired one is not available in the list).

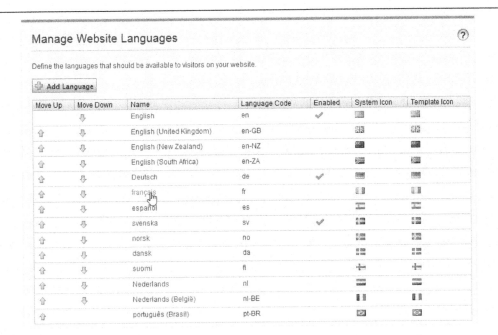

3.   Select the Enabled check box and click Save.

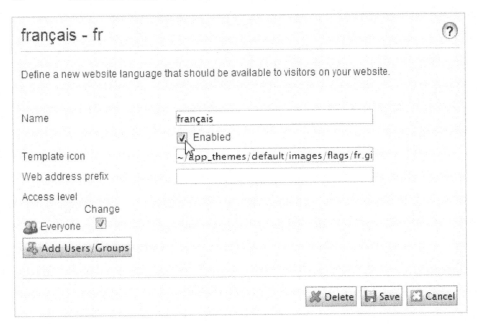

2. Enabling a language for editing in CMS

This step makes the language you just enabled available for content creation by editors. A language can be made available for the entire site structure, or for parts of it. By default, subpages inherit language settings from their parent page.

1.   In the CMS page tree, select the root page for the branch for which you want to enable the language. In this example, we want "French" to be available for the entire site, so the language setting is defined on the start page.

2.   Open the page in All Properties editing.

3.   Select Tools and Language Settings in the header.

4.   Under Available Languages, select Change.

5.   Select the language you wish to enable, click Save, and close the dialog.

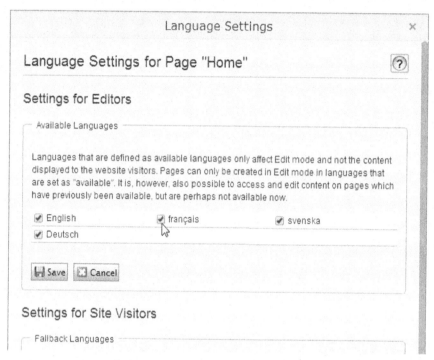

6.   For a correct display, the website start page also needs to be available in the new language. To do this, switch to the newly activated language, go to the start page, select Translate, then publish it.

Once the language has been enabled as described above, it is available for content translation.

User interface languages

The Episerver user interface is available in a number of different languages. To set the desired user interface language for CMS, click your user profile name in the upper right corner. Select **My Settings > Display Options** tab. At the **Personal Language** drop-down, select the language of your choice, and click **Save**.

Commerce and multiple languages

Refer to Multi-language management in the Commerce section for information on how to work with multiple languages if you have Episerver Commerce installed.

Find and multiple languages

Refer to Optimizing multiple sites and languages for information on how to work with multiple languages if you have Episerver Find installed.

See also

>> Refer to Translating content for information on how to translate CMS content into different languages.

>> Refer to the Languages add-on for more information on how to extend the functionality in Episerver for translating content into multiple languages.

Translating content

When you have enabled the desired language, you are ready to translate existing content, or create new content in a specific language. Content here can be, for example, pages or blocks on an Episerver CMS website, or product-related content on an e-commerce site. When a language is enabled in Episerver, content properties that are not **global** are available for translation.

 To prevent editors from accidentally creating content in the wrong language, access rights can be set differently for different languages by an administrator. If this is implemented, you can only edit and create content in languages to which you have access.

See also Languages add-on for information about extending the functionality in Episerver for translating content into multiple languages.

Switching language and viewing language versions

To switch language in **CMS**, go to the **Sites** tab in the navigation pane and select the desired language to work with. The user interface reloads, displaying the page tree in the selected language.

If your master language is English and you switch to Swedish, all pages that have not yet been translated into Swedish are displayed in italics in the page tree and with the *en* language code for English. Pages that exist in Swedish are displayed in normal font.

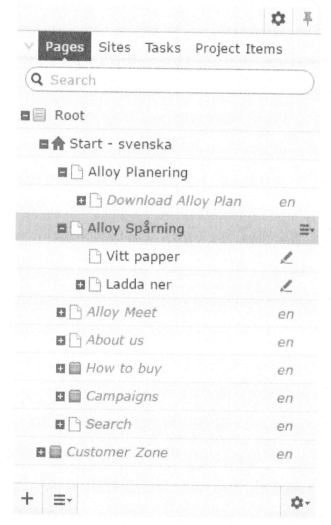

To view only pages that have been translated, select **Show Content in *[language]* Only** from the page tree's **Settings** button. This filters out all other language versions.

When you show content in one language only, you can move pages to another location in the page tree structure by drag and drop or copy and paste, but it is not possible to sort pages. Sorting is disabled since sorting in one language, where you may not see all pages, can cause unexpected result in other languages.

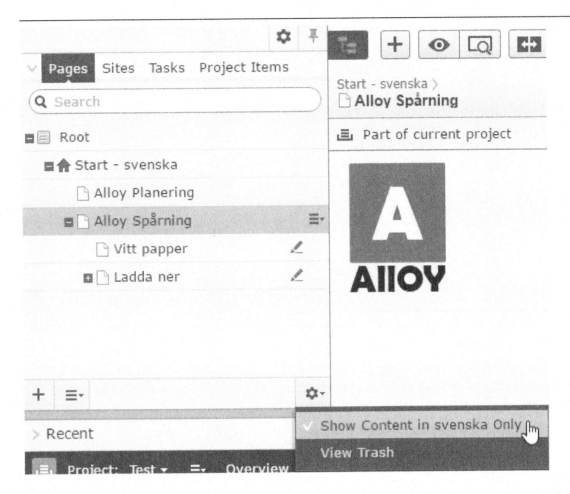

You also can switch languages by selecting the desired language in the **Header**, when editing translated content in the **All Properties** view. The user interface reloads, displaying the content in the selected language.

When you translate content, you can use the Versions gadget in the assets pane to see the different language versions for the content. By selecting a language in the version list, you also can switch to editing in another language using the switch option in the notification bar.

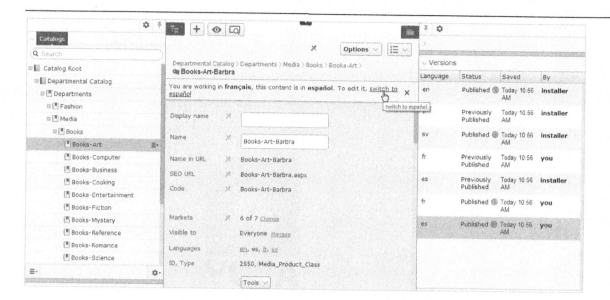

You can search for all language versions for some content by typing a keyword in the search field for **Pages** or **Blocks** in the assets pane.

Previewing content in different languages

Using the view settings in the "eye" in the top menu, you can preview and edit content in one of the languages that are available for translation on the website.

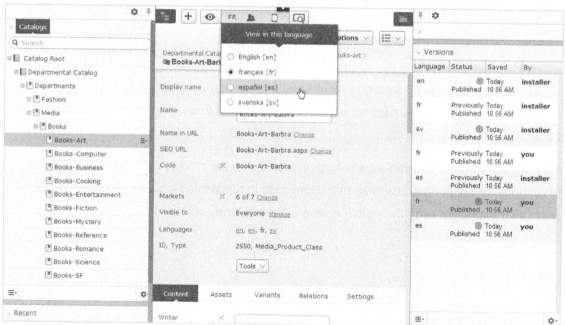

Deleting language versions

To delete one or more language versions for content, select **Delete Version** or **Delete All [language] Versions** in the **Versions** gadget.

Translating existing content

The **Sites** tab displays the languages available for content creation, with the default language for the website at the top. Languages that are enabled on the website but are not enabled for editing, are shown in italics.

Translating a page

By default, all pages in the tree structure are displayed in the **Pages** tab, including those that are not translated. These are shown in italics. To only see pages for the chosen language, select **Show Content in [language] Only**.

Do the following to translate a page:

1.   Under the Sites tab in the navigation pane, select the desired target language for translation. The interface reloads, and you are taken to the Pages tab.

2.   In the page tree, select the desired page to translate, and then click Translate in the notification bar at the top. Or, select Translate in the context menu for the page in the page tree.

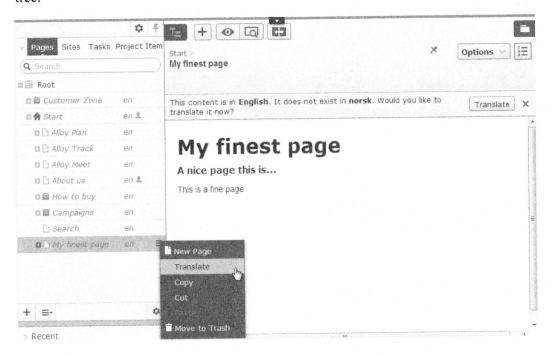

THE MISSING MANUAL

3.   You can use the compare view 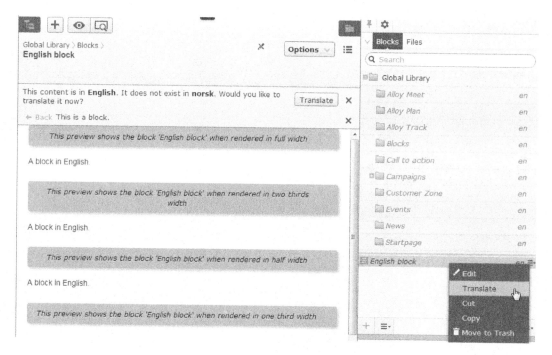 to translate in one pane while seeing the original version in the other pane at the same time.

4.   Edit the content and follow the content publishing flow to save and publish the translated page.

Translating a block

You can access language versions for blocks from **Blocks** in the assets pane. By default, all blocks are displayed, including those that are not translated; these are shown in italics. To only see blocks for the chosen language, select **Show Content in [language] Only**.

To translate a block:

1. Under the Sites tab in the navigation pane, select the desired target language for translation, and the interface reloads.

2. Expand the assets pane and select Blocks.

3. In the block structure, select the desired block to translate, and then the Translate option in the context menu.

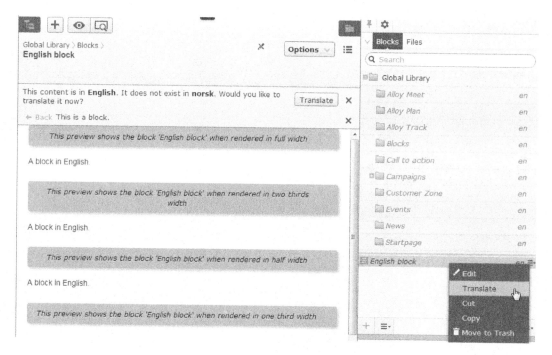

4. You can use the compare view to translate in one pane while seeing the original version in the other pane at the same time.

5. Edit the content and follow the content publishing flow to save and publish the translatedblock.

© Episerver

Global properties

Depending on your implementation, some properties may be "globally shared "and you can edit them in the master language. These properties are marked as **non-editable** when editing the content in another language. Switch to the default language if you need to edit these. The default language is usually the first language listed next to **Languages** in the header when editing in the All Properties view.

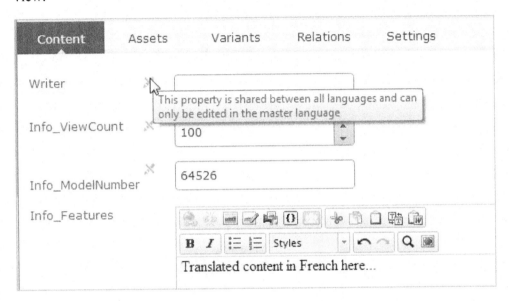

Creating new content in another language

To create content in a language other than the "master "language, select the desired language under the **Sites** tab. Then select the page branch or folder where you want to create the content, and create a new page or a new block. Add content and follow the content publishing flow to save and publish.

Commerce **Commerce and multiple languages**

See Multi-language management in the Commerce user guide for information about working with multiple languages if you have Episerver Commerce installed.

CMS **Fallback languages**

For many multi-language websites, only parts of the website content exist in all available languages. The reason can be that translations are not yet ready, content is not relevant fora specific language, or that some content should always display in a defined language.

You have the following options:

» Unless a fallback or replacement language is defined, content is invisible to visitors browsing the website in a language into which content is not translated.

» Define a fallback language, in which the content is displayed until the content is available in the desired language.

» Define a replacement language, in which content is always displayed regardless of the language in which the content exists. If you define a replacement language for some content, a fallback language does not apply.

 Fallback and replacement languages may cause mixed languages to be displayed on the website.

Setting fallback and replacement languages

Fallback and replacement languages are defined from the **All Properties** edit view, when editing a page or a block and then selecting **Tools** and **Language Settings**.

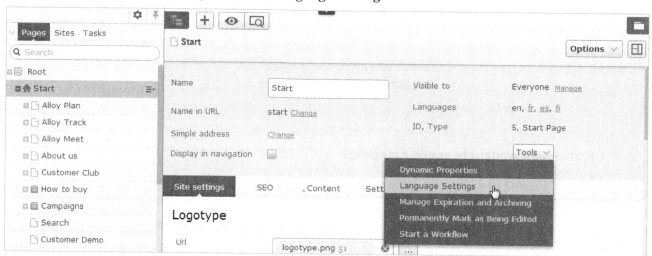

 Language settings are inherited from the parent page. If you are redefining language settings fo a subpage to a parent page with language settings defined, you need to deselect **Inherit settings from the parent page "xxx"** in the **Language Settings** dialog, to define settings for the subsection.

EXAMPLE: fallback language

In this example, the *master* website language is *English*, and *Swedish*, *Danish* and *Norwegian* are enabled languages. Content is first created in *English*, and then translated into *Swedish*,

Norwegian and *Danish* in that order. Swedish is used as first fallback for *Norwegian* and *Danish*. If content does not exist in *Swedish* (not translated yet), then a second fallback language *English* is applied.

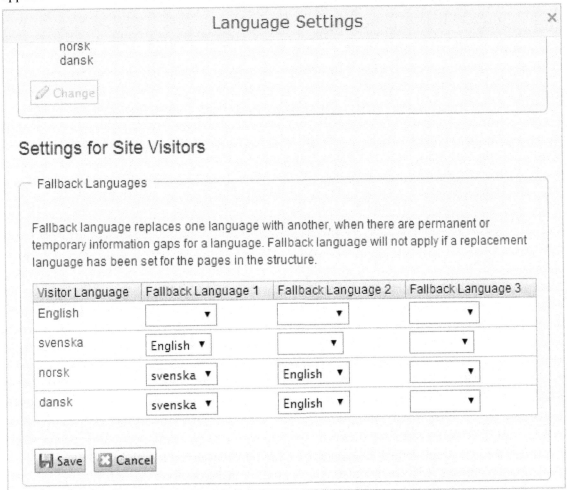

EXAMPLE: replacement language

This example shows a multi-language website with a legal information section with content that always should be displayed in English. To ensure this, a replacement language is applied for the legal information page branch.

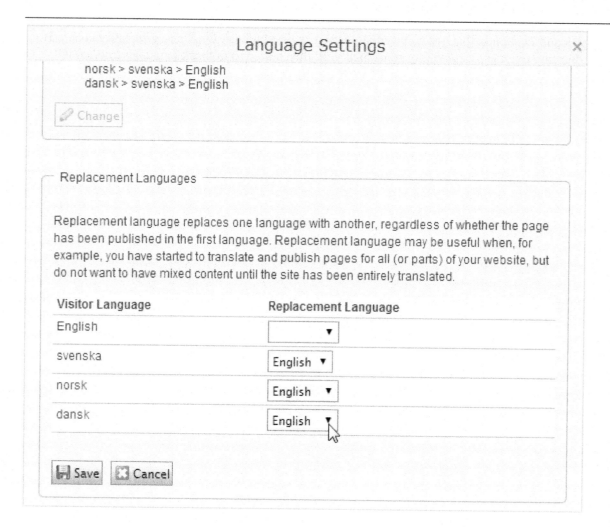

Another scenario for using replacement language is when you have started to translate content on your website, but do not want to have mixed languages until translation is completed. When translation is ready, you simply remove the replacement language.

See also

» Translating content for information about translating CMS content into different languages.

» Languages add-on in the online user guide for information about extending the functionality in Episerver for translating content into multiple languages.

Personalizing content

Personalization in Episerver lets you target website content for selected visitor groups. For example, you can design a product banner, a landing page or a registration form specifically for first-time visitors, or for visitors from a geographic region or market.

The personalization feature is based on customized **visitor groups**, which you create using a set of **visitor group criteria**. Visitor groups must first be created to become available for selection when applying personalization.

There are numerous visitor group criteria available across the Episerver platform; see Administering visitor groups in the CMS Administrator User Guide.

Working with personalization

You can personalize any type of content in the rich-text editor and in a content area. Personalize part of a text, an image, or a block in the rich-text editor, or personalize an image, a block or a page in a content area, if you have these in your web pages.

If you have multiple visitor groups, a visitor may match more than one visitor group. You then can use **personalization groups** to group content to avoid displaying the same content twice, and display **fallback content** to visitors who match no visitor groups.

The preview option in the top menu lets you preview the personalized content as the different visitor groups will see it, before publishing.

Applying personalization

In the rich-text editor

1. Open the page for editing, and select the content you want to personalize in the editor area.
2. Click Personalized Content 🖼 in the editor toolbar.

3. Select one or more visitor groups from the list.

CMS Search

The Episerver platform has sophisticated search functionality that lets you search through different types of website content. You can search for content pages, blocks, files, community objects and products, if Episerver Commerce is installed. The search results are based on access rights, so users only see content to which they have access.

The search is based on the open-source search engine Lucene, which is used by the different Episerver products when retrieving content. The search is provider-based, letting you extend and customize features. You can use the built-in basic search features or create your own custom filtering methods and queries.

Built-in search features

The following built-in search features are included in Episerver:

>> Full-text search features.

>> Indexing of all content types: document files, pages and blocks.

>> Event-driven indexing, meaning instant updating of the index and search results.

>> Search results filtered on access rights.

>> "Search-as-you-type," enhancing the search experience.

>> Supports basic boolean operators like AND, OR, NOT, +, - and *.

Search options

Depending on how your Episerver installation is set up and from where you are searching, there are different options:

>> When editing, the global search is available in the upper right part of the global menu. Depending on the configured search providers, this option can search all types of content on the website—pages, blocks, files, and catalog content if you have Episerver Commerce installed.

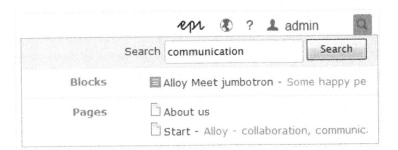

》 When editing, a search option is available at the top of the navigation and assets panes, and in the link dialog. This option searches for content in the panes and related dialogs.

Search tips

》 Enter a few carefully selected keywords separated by space. If need be, narrow down your search query by adding keywords. For example: EPiServer product project.

》 If you know a page ID, search for the page by entering the ID in the search field.

》 When searching for specific phrases, combine keywords using quotation marks. Example: "EPiServer search tips".

》 The search function is case-insensitive, so you can use both uppercase and lowercase letters. Example: New York and new York return the same result.

》 You can restrict the search by placing a plus sign (+) in front of the words that must be found to consider the page a match. Example: +EPiServer +search +tips.

》 Similarly, you can restrict the search by placing a minus sign – in front of the words that must not occur to consider the page a match, for example -EPiServer -search -tips.

》 To match part of a word, place an asterisk * at the end of the word. Example: word1* word2 return content with the words word10, word123 and word2, but not word234.

》 You can use the boolean operators AND and OR.

>> AND means I only want documents that contain both/allwords.Example: episerver AND search returns documents with both words.

>> OR means I want documents that contain either word. Example: episerver OR search returns documents with either EPiServer or search.

Sorting search results

The sorting of search results is determined by the search algorithm, which you can customize in many ways. Often, filtering is applied to the results, which can be based on many factors, such as categorization of content.

Configuring search

Episerver search has configuration options that are managed from the administrative interface in Episerver CMS. See EPiServer's technical documentation for information about search functionality, configuration possibilities, and integration interface.

Addons **Extended search with EpiserverFind**

To build more advanced search features based on visitor behavior, customized filtering and faceted content navigation, you can add Episerver Find to your solution (requires license activation); see the Episerver Find User Guide.

EPISERVER
ADMINISTRATOR GUIDE

THE MISSING MANUAL

Introduction

The features and functionality of the entire Episerver platform are described in an online help that opens in a web browser. The online help covers CMS for content management, Commerce for e-commerce functionality, Find for extended search, and Episerver add-ons. It is either accessed from within the Episerver platform or from Episerver World. The online help is also divided into a number of PDFs for users who prefer those or want to print the documentation.

This PDF describes the features and functionality of Episerver CMS. PDFs for Episerver Commerce and Find can be found on Episerver World. The user documentation is intended for editors, administrators, marketers and merchandisers, working with tasks as described in Roles and tasks.

Developer guides and technical documentation are also found on Episerver World.

Features, licenses and releases

The user documentation is continuously updated and covers the latest releases for the Episerver platform.

CMS   Episerver CMS is the core part of the Episerver platform providing advanced content creation and publishing features for all types of website content. CMS features are available in all Episerver installations.

Commerce   Episerver Commerce adds complete e-commerce capabilities to the core functionality in CMS. Commerce requires additional license activation.

Addons Add-ons extend the Episerver capabilities with features like advanced search, multivariate testing, and social media integration. Some add-ons are free, others require

license activation. Add-ons by Episerver are described in the online help.

 Due to frequent feature releases, this user guide may describe functionality that is not yet available on your website. Refer to What's new to find out in which area and release a specific feature became available.

Copyright notice

What's new?

The Episerver user guide describes features in the Episerver platform, including CMS for content management and Commerce for e-commerce management, and add-ons from Episerver. New features are continuously made available through Episerver updates.

This user guide (16-6) describes **features added up until and including update 127** for

Episerver; see_ Episerver World for previous user guide versions.

Area		Features and updates
CMS	» »	Access rights was revised to expand its information.
	»	You can watch the following demonstration video, Managing access rights. (6:39 minutes) You can watch the following demonstration video, Publishing content. (4:18 minutes)
Commerce	»	A new discount lets you give free shipping to customers who buy a minimum number of items. See Buy products for free shipping. (update 125)
	»	A new discount lets you give free shipping to customers who spend a minimum amount on an order. See Spend for free shipping. (update 125)
	»	A new discount lets you give a discount on every item to customers who order a minimum number of items. See Buy products, get discount on all selected. (update 126)
	»	A new discount lets you give the most expensive item for free to customers who order a minimum number of items. See Get most expensive for free. (update 126)
	»	You can watch the following Demo of creating a campaign and iscount video. (4:43 minutes)
Addons	»	

Episerver Forms has a Marketing Automation Integration connector that lets you connect
Episerver form fields to a Digital Experience Hub (DXH) connector database.

Getting started

This section describes how to log in to an Episerver website, access features and navigate the different views. Note that the login procedure may be different from what is described here, depending on how your website and infrastructure are set up. The examples described here are based on a "standard" installation of Episerver with sample templates.

Logging in

As an editor or administrator, you usually log in to your website using a specified URL, a login button or link. Enter your user name and password in the Episerver login dialog, and click **Log In**.

Accessing features

What you are allowed to do after logging in depends on your implementation and your access rights, since these control the options you see. When logged in, the Episerver quick access menu is displayed in the upper right corner.

Selecting **CMS Edit** takes you to the edit view as well as other parts of the system. You can go directly to your personal dashboard by selecting the **Dashboard** option.

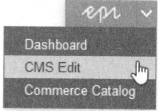

Navigation

Pull down the **global menu**, available at the very top, to navigate around. The menu displays the different products and systems integrated with your website. Select, for instance, **CMS** to display available options in the submenu.

Next steps

Refer to the sections below for more information.

» User interface and Roles and tasks in the CMS Editor User Guide for information about the Episerver user interface and roles.

» Managing content in the CMS Editor User Guide for information on how to create and publish content.

» Administration interface in the CMS Administrator User Guide for information on how to administer and configure settings in Episerver.

» Commerce User Guide for information on how to work with e-commerce tasks, if you have Episerver Commerce installed.

» Find User Guide for information on how to work with search optimization, if you have Episerver Find installed.

» Add-ons section in the online help for information on how to use add-ons from Episerver, if you have any of these installed.

Roles and tasks

Episerver is designed for interaction with website visitors, as well as collaboration between users. A user in Episerver is someone working with different parts of the platform. A user can belong to one or more user groups and roles, depending on their tasks as well as the size and setup of the organization.

Typical roles and related tasks are described below. Refer to Setting access rights in the CMS Administrator User Guide for information on how to configure user groups and roles in Episerver.

Visitor

A visitor is someone who visits the website to find information or to use available services, on an ecommerce website possibly with purchasing intentions. Purchasing on an e-commerce website can be done either "anonymously" (payment and shipping details provided), or by registering an account. Visitors may also contribute to website content as community members, which usually requires registration of an account profile.

Community member

Content may be added by visitors or community members, if social features and community functionality are available for the website. This content includes forum and blog postings, reviews, ratings and comments, in which case there might be a need for monitoring this type of content on the website. Monitoring can be done for instance by an editor, or a specific moderator role for large websites and online communities.

Content editor

A content editor is someone with access to the editorial interface who creates and publishes content on the website. Content editors with good knowledge of the website content work with search optimization for selected content in search results. Editors may also want to follow-up on content with unusually high or low conversion rate in order to update or delete this content.

Marketer

A marketer creates content and campaigns with targeted banner advertisements to ensure customers have consistent on site experience of the various marketing channels. Furthermore, the marketer monitors campaign KPIs to optimize page conversion. A marketer with good knowledge of the website content may also want to monitor search statistics in order to optimize search for campaigns and promote content.

Merchandiser

A merchandiser typically works with stock on an e-commerce website to ensure that the strongest products are put in focus. This role also creates landing pages, sets product pricing, coordinates cross product selling, oversees delivery and distribution of stock, and deals with suppliers. This user wants to be able to identify search queries with unusually high or low conversion rates, in order to adjust either the search or the product line.

Website owner

A website owner is someone with overall responsibility for the content and performance of one or more websites. This user monitors website activities such as page conversions, customer reviews or sales progress. Rarely creates content but can be involved in the approval of content created by others. A website owner may have administrative access rights and may be able to install selected add-ons on the website.

Administrator

An administrator works with configuration of various system settings from the administration user interface, including search, languages, user access rights and visitor groups for personalized content. Administrators may also install add-ons on the website. Administrators usually have extended access rights compared to other user groups, and can access all parts of the Episerver platform.

Developer

A developer is someone with programming skills working with the setup and implementation of the website, as well as maintenance and development of new functionality. Creates the content templates for pages, blocks and catalog content used by editors in CMS and Commerce, configures e-commerce Getting started settings, and manages the index and customized search features in Find. Developers may also install add-ons on the website.

CMS Administration interface

Depending on which parts of the Episerver platform are implemented, you have various administration options in the user interface. The options in this topic apply to a standard installation of Episerver and related products; a customized site might have additional administration options.

Administration view

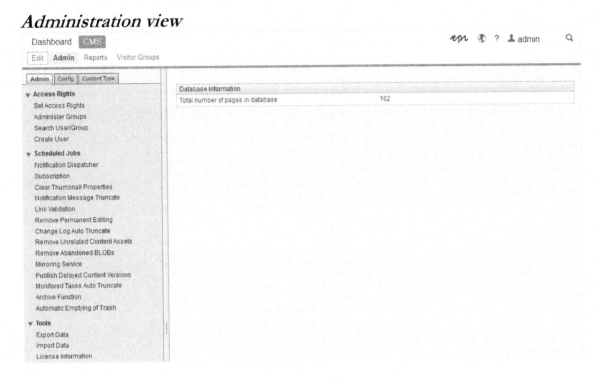

The **admin view** in CMS contains core administration features for the Episerver platform where you manage access rights, **website** languages, **and** scheduled jobs. You also manage the export and import of data between websites, and configure new websites in a multi-site solution.

Visitor groups

Visitor groups are used by the personalization feature, and are managed from the **Visitor Groups** option in the global menu. You need administration access rights to manage visitor groups.

More on administration

Commerce Administration in Commerce

If you have Episerver Commerce installed on your website, Commerce has an administrative interface for managing e-commerce-specific settings. See the Administration section in the user guide for Episerver Commerce.

Addons Administration in Find

If you have added Episerver Find to your website, there are some specific administration and configuration options available to optimize the search functionality. See the Administration and configuration section in the user guide for Episerver Find.

Access rights

You can control the content that a visitor sees and the content that users can edit on your website by setting access rights on content such as pages, blocks, media, and folders. A user or group has access rights on a per-content basis. For example, you may give members of the Marketing department access to edit the main website marketing material that other company users should not edit. Or you may want to give a visitor group from a local 10-mile radius access to an advertizement page.

 You can watch the following demonstration video, Managing access rights. (6:39 minutes)

You can define specific access rights from the "Root" level and all the way down, including the **Recycle bin** (Trash) and **For All Sites** that stores blocks and media.

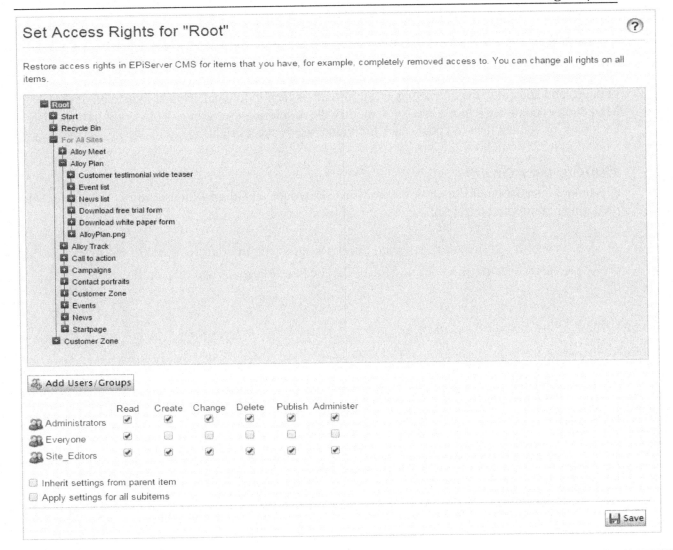

Set Access Rights for "Root"

Restore access rights in EPiServer CMS for items that you have, for example, completely removed access to. You can change all rights on all items.

Blocks and media share the same folder structure. If you want to automatically publish media that are uploaded to the website , editors who upload must have **Publish** access rights in the folder (under **For All Sites**) to which the media are uploaded. Also, editors must have **Create** access rights in the **root** level of the website to create blocks.

You can set access rights to content for a single user. For example, you can make it so only *Ann* (and system administrators) can edit the *Book a Demo* page. You can add *Ann* to any number of pages and content, and set *Ann's* access rights to each content item the same (or differently) for each page.

If you have a number of users that should have common access to content, managing access rights on a user-by-user basis can become complex. You should create user groups that have similar access needs, add the users to each user group, and then use the user group to set access rights to content. This lets you manage access rights more easily. You can add a user to one or more groups.

For example, add *Ann*, *Bob*, and *Cam* to a *Marketing* user group and give access rights to any number of pages and content to the *Marketing* group instead of each individual. To add *Dan* to all of the Marketing content, (or remove *Ann*), you modify the *Marketing* user group. You do not have to visit each page or content item to update each individual user's access rights.

Built-in user Groups

A standard installation of Episerver has built-in user groups that align with user roles. You can extend predefined groups and roles; see Managing users and user groups

 When your website is set up during development, configure the membership and role providers available for your website to use the built-in groups and roles in Episerver.

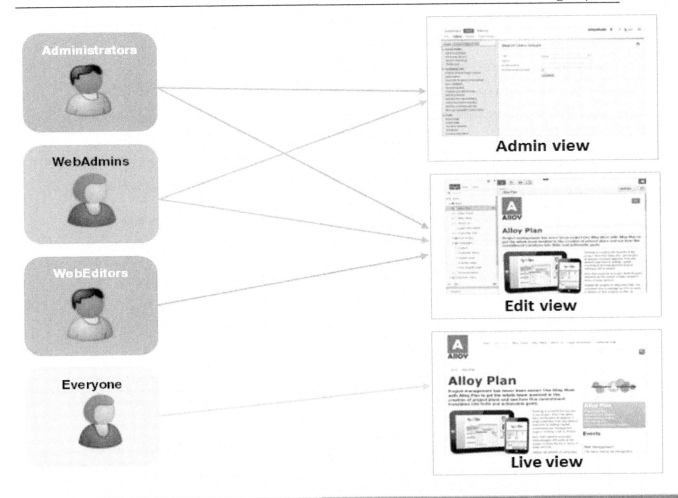

Group	Description
Administrators	Comes from Windows and is defined when the website is created, an admin - istrator can access all parts of the system, and can edit all website content. Often,

Setting access rights

1. Go to CMS > Admin > Admin tab > Set Access Rights. The Set Access view appears with a content tree structure of the website.

2. Click on a node in the content tree (for example, Marketing). Typically, a content item shows Administrators (with all access rights) and Everyone (with Read only access rights). You can change these rights or add new users or groups.

   » If the users or groups are inactive (grayed out) for a content item, then the content item inherits the access rights of its parent content item. To set access rights for this content item, clear the Inherit settings from parent item check box. You can modify the access right for existing users or groups or add new ones.

>> To add settings to all subitems without affecting their existing settings, select the Apply settings for all subitems; see example.

3. Click Add Users/Groups. A dialog box appears.

4. Select the type you want: Users, Groups, or Visitor Groups.

5. Leave the Name field blank and click search to display all items of the type you selected. You can also type one or more characters in the Name field to filter and display a subset of items.

   (You also can select a user by E-mail address.) For example, add the Marketing Group group and user Ann to the Marketing content item's access rights.

6. Double click a user or group in the Existing box to move it to the Add box and click OK.

7. Modify the access rights settings as you want them and click OK. The users or groups appear in the Set Access Rights view for the select content tree item.

You manage access rights from the administration view in Episerver CMS, but you also can let editors manage access rights for a single page in edit view.

 If you set conflicting access rights to content, selected access rights prevail over cleared access rights. For example, *Ann* is a member of both the *Marketing* and *Support* user groups which each have different access rights set on the same content. (Perhaps *Marketing* has **Publish** rights, but *Support* does not.) *Ann*, who is in both groups, has **Publish** rights to the content, but *Bob*, who is part of the *Support* group only, does not have **Publish** rights.

Setting in heritance for content subitems

Content inherits access rights from its closest parent item. When you set access rights for a content item, the rights apply to it and all subitems that have a selected **Inherit settings from parent item** option; subitems with this option cleared are not affected. For example, the following content items all have the same access rights because Alloy Meet, Alloy Plan, and Alloy Track inherit the access rights from the Alloy content item.

Alloy
  Alloy Meet
  Alloy Plan
  Alloy Track

>> If you break inheritance for Alloy Meet and give access to user Ann, Bob, and Cam, the access rights become different from the parent (Alloy) and its two siblings (Alloy Plan and Alloy Meet). >> If you then break inheritance for Alloy (parent) and add a Marketing group. Alloy Plan and Allow Track inherit the Marketing group (because inheritance is selected) but Alloy Meet does not because its inheritance is unchecked.

Subitems inherit from the closest parent only if the inheritance option is active (checked).

The following image shows the access rights for the **Marketing** content; no inheritance is set from the parent item or for subitems.

Product A is a subitem of **Marketing**. It has **Inherit settings from parent item** selected, so the access rights are identical to that of the **Marketing** content item.

Book a demo has **Inherit settings from parent item** cleared, so its access rights are not the same at the **Marketing** parent content item. (**Ann** does not show up, and **Zach** is listed in the access rights.)

Applying settings for all subitems
Apply settings for all subitems applies the access rights of the parent item to all its subitems, even if a subitem has inheritance cleared. The option adds settings to a subitem that it did not have before and does not or remove any settings that the subitem already had.

For example, the **Marketing** content item has **Ann** as a user with access rights.

When you **Apply settings for all subitems**, **Ann** is added as a user with access rights to **Book a demo** because **Ann** is part of the **Marketing** content item's access rights. However, **Zach** remained on the list of access rights, unchanged.

If a parent item has a user or group that is the same as a user or group in a non-inheriting subitem (but each item has different access rights for the user or group), when you select **Apply Settings for all subitems**, the parent settings are applied to the subitem. For example:

>> If the Marketing parent item has user Ann with only Read access set, while the Book a Demo subitem also has user Ann but with all access rights, then Apply Settings for all subitems resets the access rights for Ann on the Book a Demo subitem to only Read access. >> Conversely, if Marketing has user Ann with all access rights, and the subitem

THE MISSING MANUAL

has user Ann with only Read access, Apply Settings for all subitems gives user Ann all access rights on the subitem.

Removing a user or group from the access rights list

To remove a user or group from the access list, clear all of the access rights for that user or group and click **Save**.

Using a visitor group in an access rights list

Visitor groups are used by the personalization feature, and you need administration access rights to manage visitor groups. If you want an editor to manage visitor groups without providing access to the entire admin view, you can make the editor a member of **VisitorGroupAdmins**. This group provides access only to the **Visitor Groups** option in the global menu. **VisitorGroupAdmins** comes with Episerver but you must add this group name through **CMS > Admin view > Admin tab > Administer Groups > Add > VisitorGroupAdmins**.

You can set specific access rights for visitor groups, letting them view specific "hidden" content that is not otherwise publicly available. For example, you may want only members of the *Visitors from London* visitor group to have access to a *Family day at the zoo* page with a discount coupon.

This feature is useful if you want to create a "customer area" for registered customers on your website. Being a member of a visitor group requires a registration and login to access the content.

1. Go to CMS > Admin > Admin tab > Access Rights > Set Access Rights.
2. Click Add Users/Groups. The Add Users/Groups dialog box appears.
3. Select the Visitor groups type and select a visitor group. (Click Search while leaving the Name field blank to view available visitor groups.)

Access rights for Add-ons

Add-ons are plug-ins for extending Episerver functionality. You need administration access rights to manage add-ons.

has user Ann with only Read access, Apply Settings for all subitems gives user Ann all access rights on the subitem.

Removing a user or group from the access rights list

To remove a user or group from the access list, clear all of the access rights for that user or group and click **Save**.

Using a visitor group in an access rights list

Visitor groups are used by the personalization feature, and you need administration access rights to manage visitor groups. If you want an editor to manage visitor groups without providing access to the entire admin view, you can make the editor a member of **VisitorGroupAdmins**. This group provides access only to the **Visitor Groups** option in the global menu. **VisitorGroupAdmins** comes with Episerver but you must add this group name through **CMS > Admin view > Admin tab > Administer Groups > Add > VisitorGroupAdmins**.

You can set specific access rights for visitor groups, letting them view specific "hidden" content that is not otherwise publicly available. For example, you may want only members of the *Visitors from London* visitor group to have access to a *Family day at the zoo* page with a discount coupon.

This feature is useful if you want to create a "customer area" for registered customers on your website. Being a member of a visitor group requires a registration and login to access the content.

4. Go to CMS > Admin > Admin tab > Access Rights > Set Access Rights.
5. Click Add Users/Groups. The Add Users/Groups dialog box appears.
6. Select the Visitor groups type and select a visitor group. (Click Search while leaving the Name field blank to view available visitor groups.)

Access rights for Add-ons

Add-ons are plug-ins for extending Episerver functionality. You need administration access rights to manage add-ons.

If you want an editor to manage add-ons without providing access to the admin view, make the editor a member of **Packaging Admins** group which provides access only to the **Add-ons** option in the global menu.

PackagingAdmins comes with Episerver but you must add this group name through

 CMS > Admin view > Admin tab > Administer Groups > Add > PackagingAdmins.

Some add-ons also may have specific user groups defined to access the functionality. See the documentation for each add-on in the online user guide to find out more.

Access rights for languages

If your website has content in multiple languages, you can define access rights for languages so editors can create content only in languages to which they have access. Only users with access rights for a language have it available on the **Sites** tab, and can create and edit content in that language. See Configuring website languages.

Commerce ## Access rights for Commerce

If you have Episerver Commerce installed on your website, see the Commerce access rights section in the Commerce User Guide .

Addons ## Access rights for Find

If you have added Episerver Find to your website, see the Find access rights section in the Find User Guide .

CMS ## Managing users and user groups

For easier and safer maintenance, you should base access rights on **user groups** rather than individual users.

You can administer user credentials in the following ways:

>> Manage users and user groups from the CMS administration view.

>> Manage users and user groups in Windows.

>> Develop a customized role and membership provider.

Smaller organizations with few editors tend to use the CMS administration view, whereas larger organizations with many editors tend to use the other options. You can combine these options.

Creating, editing and deleting users

To add a user in the CMS, do the following:

1. On the Admin tab, select Create User.

2. Specify a username, password and email address and set the account to Active.

3. Select none or more user groups to which the user should belong and click the arrow (or double click) to place the selected groups in the Member of box.

4. Under the Display Options tab, you can specify a default language for the user interface (optional) and touch support.

5. Click   Save.

6. To **edit** user settings, search for the user under **Search User/Group** and then click a user name. You can modify properties only for users that are created via self-registration or via **Create User** in CMS.

7. To **delete** a user, search for the user under **Search User/Group**, click a user name to edit the settings, and click **Delete**. You cannot undo a deletion of a user.

8. **Displaying members of a user group**

10. Select the **Search User/Group** option to display groups and users.

11. Group view. Click the desired group name to see members in the group.

>> User view. Click on a user name to display the Edit User panel where you can modify group memberships and other user settings.

>> To see all users or groups, leave the Name field blank.

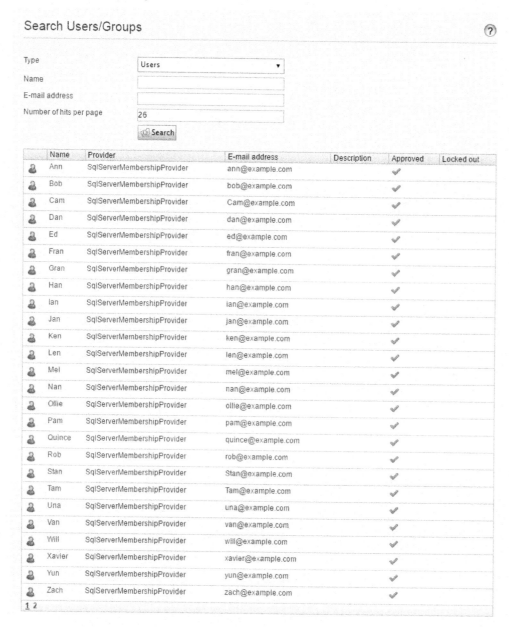

>> To see a subset of users or groups, type one or more letters in the Name field. Any names that have the string in them appear in the list.

CMS   Scheduled jobs

A **scheduled job** is a service performing a task (job) at a given time interval. An administrator can start a job manually. A standard installation of the Episerver platform with Episerver CMS and Episerver Commerce includes several scheduled jobs. Some are enabled by default with preset values. You can develop customized scheduled jobs for specific website tasks.

Administering scheduled jobs

Manage scheduled jobs as follows:

1. Log in as an administrator and go to the Episerver CMS admin view.

2. Select the desired scheduled job on the Admin tab > Scheduled Jobs.

3. Select the Active box to activate the scheduled job.

    » To run the scheduled job automatically, set the desired time interval in

    Scheduled job interval. Each scheduled job's run time appears in the Next

    scheduled date field. » To run the scheduled job manually, click Start Manually

    and the job is executed immediately.

4. Click Save.

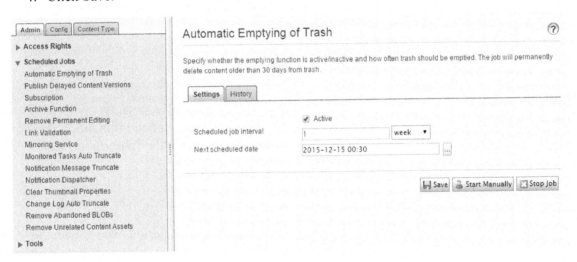

The **History** tab lets you monitor the status and results when a scheduled job is executed. If a job fails, information about it appears under the **Message** column.

Date	Status	Message
5/21/2013 10:34:54 AM	OK	0 content items were deleted from recycle bin.
5/20/2013 8:55:01 AM	OK	0 content items were deleted from recycle bin.

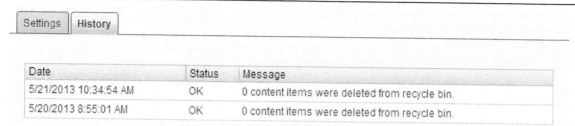

Built-in scheduled jobs

Automatic Emptying of Trash

You can set up how often your trash gets emptied with the **Automatic Emptying of Trash** job. With automatic emptying, all content in trash older than 30 days is permanently deleted by default. Trash also can be permanently deleted manually.

The job is enabled by default, and set to run weekly.

Publish Delayed Content Versions

The **Publish Delayed Content Versions** job lets you define how often the system checks for content versions with a specific future publication date and time.

The job is enabled by default, and set to run hourly.

Subscription

The **Subscription** feature lets visitors set the frequency for receiving subscription information. This job checks for information from the system to be included and distributed in the subscription send-out.

Archive Function

You can set how often the system archives information after the publication period expires with the **Archive Function** job.

 There can be a delay between the time information is unpublished, and when it appears in the archive. This may occur if the archiving job is run only once a day.

Remove Permanent Editing

You can clear the **Permanently Mark as Being Edited** marking of pages in the edit view (if editors have forgotten to remove the marking) with the **Remove Permanent Editing** job.

The job is enabled by default, and set to run hourly.

Link Validation

You can check links on your website to identify broken links with the **Link Validation** job. The system tries to contact the target for the link to verify that it is responding.

Links are returned only if they are unchecked or checked earlier than the time when the job started. The job continues until no more unchecked links are received from the database. If a large number of consecutive errors is found for external links, in case of a general network problem with the server running the site, the job stops.

The result of the link validation job is a report called **Link Status**, in the **Episerver CMS Report Center**.

Mirroring Service

You can set the frequency of mirroring content between websites with the **Mirroring Service** job. If your website is set up to mirror content between websites, you can manually mirror content or automatically do so at specific intervals. See also Mirroring.

Monitored Tasks Auto Truncate

The **Monitor Tasks Auto Truncate job** truncates the status of monitored tasks. It is a clean-up job that deletes 30 days of statuses of monitored and completed jobs.

The job is enabled by default, and set to run weekly.

Notification Message Truncate

The **Notification Message Truncate** job truncates or deletes 3 months old notification messages that could not be sent and are still in the system.

The job is enabled by default, and set to run every day.

Notification Dispatcher

Set the **Notification Dispatcher** job to determine how often Episerver CMS sends notifications of new or updated comments or replies posted in projects by a notification provider (for example, an email provider). Notification messages are sent to:

>> users who are tagged in a comment or reply >> users who receive replies to their comments

>> users who receive comments on their project actions (such as setting a project item to

Ready to publish)

>> other users who have previously replied to the same comment

A notification is not sent if no new comments or replies were posted since the job last executed.

The job is enabled by default, and set to run every half hour.

Clear Thumbnail Properties

You can clear generated thumbnail images in the Products list and Media list views and add them again with the **Clear Thumbnail Properties** job. Run this job manually if you experience problems with refreshing thumbnails, such as on the website and BLOB-supported content.

Change Log Auto Truncate

You can delete items from the change log that are more one month old and do not have any dependencies registered against them by another part of Episerver CMS (for example, Mirroring) with the **Change Log Auto Truncate** job.

The job is enabled by default, and set to run weekly.

Remove Abandoned BLOBs

Episerver CMS can store media files in a cloud service instead of the website's database. When you delete CMS files, this job makes sure the stored data is deleted from the BLOB provider.

The job is enabled by default, and set to run weekly.

Remove Unrelated Content Assets

You can delete content folders that contain media related to deleted content items with the **Remove Unrelated Content Assets** job.

The job is enabled by default, and set to run weekly.

`Commerce` *Commerce-related scheduled jobs*

Installing Episerver Commerce adds scheduled jobs to your implementation. See Scheduled jobs in the Commerce user guide for information.

`Addons` *Find-related scheduled jobs*

See Administration and configuration in the Episerver Find user guide for information about scheduled jobs for Find.

Other scheduled jobs

Customized modules and add-ons may have their own specific scheduled jobs. See the technical documentation for each module to find out more.

CMS Exporting and importing data

You can export and import data between Episerver websites. This function is widely used by developers building new functionality in a test/development environment. When you complete work and the information is ready for the production environment, use the export and import features to transfer the data between websites.

Exporting data

You can export the following:

- » Content items
- » Content types
- » Frames
- » Dynamic property definitions
- » Tabs
- » Categories
- » Files
- » Visitor groups

When you select a type of item to export, available items of that type on the website are displayed.

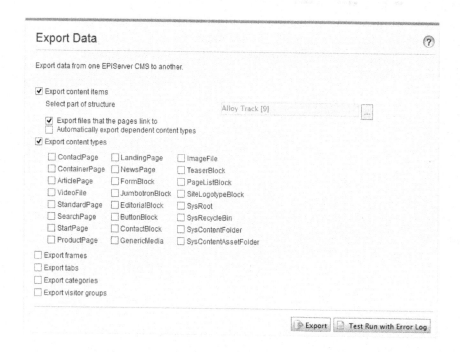

Select the items to transfer and click **Export** to download the file package.

Importing data

With the **Import Data** function, you can retrieve information exported from another Episerver website. Start by selecting the file package to import, files must end with *.episerverdata for the import to work.

Click **Upload and Verify File** to verify the file content. The files are read and checked, and verification information is displayed. Select a destination to add imported pages, and click **Begin Import**.

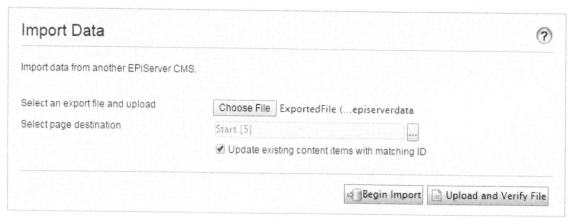

If you select **Update existing pages with matching ID** check box, the import keeps the same GUID based identities for items (such as pages, blocks and files) as they had on the exporting site. The import checks whether an item already exists and, if true, that item is updated (if the imported item had a changed date that is later than the existing item). This means that content items with the same ID are replaced instead of added, with every new import.

CMS # System settings

System settings let you define certain settings for the Episerver CMS installation, for instance, to activate globalization, change the error handling, and configure version management of content.

General tab

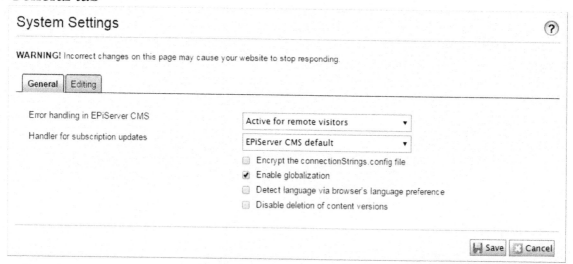

Setting	Description
Error handling in Episerver CMS	Select how you want errors to be handled; active for all visitors, remote visitors, or disabled.
Handler for subscription updates	The subscription function in Episerver lets visitors receive information about new and updated pages. Depending on whether multi-language is supported, you can select how the subscription dispatch is managed. This list also can include your own solutions for the subscription function.
Encrypt the con- nectionStrings.config file	Select to encrypt the **connectionStrings.config** file, which contains sensitive information sent to and from the database.
Enable globalization	Select to activate management of content in multiple languages (glob- alization).
Detect language via browser's language pref- erence	Select to activate languages to be shown based upon the visitor's browser settings.

Editing tab

System Settings

(?)

WARNING! Incorrect changes on this page may cause your website to stop responding.

| General | Editing |

| Path to CSS file for the Editor | `~/Static/css/Editor.css` |
| Maximum number of versions | `20` |

☐ Unlimited versions
☑ Automatically publish media on upload
☑ Enable Projects

[💾 Save] [✖ Cancel]

Setting	Description
Path to CSS file for the rich-text editor	Controls the appearance of the rich-text editor. This can be the same or similar CSS file as the site uses for styling content so that the editors get the same appearance as the site when editing content. You also can set other CSS files for different editors on the website. This is a dynamic property that you can change in the edit view.
Maximum number of versions	Specify the number of previously-published versions of content items (for example, pages or blocks) that are stored. The currently-published version and draft versions are not counted. For example, if you enter **3**, Episerver CMS stores three previously-published versions. If that is the case for a content item and you publish a new version, the oldest version is removed. Default value is 20 versions. This field is ignored if the **Unlimited versions** or **Disable deletion of content versions** box is selected.
Unlimited versions	Stores an unlimited number of versions of content items (such as pages or blocks). This option may result in a large version list, which can be difficult to

CMS   Managing websites

You can easily add and remove websites from an Episerver installation, perhaps to create short-lived campaign websites.

On the **Config** tab > **Manage Websites**, you can see an overview of existing websites in your installation. These websites share the same database, content types and templates, making it easy to set up new websites. You also can define whether content, such as blocks and folders, should be shared or site-specific.

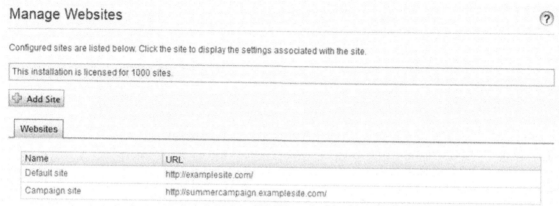

You can add new websites in the following ways:

 >> Single-site setup lets your installation have one CMS site mapped to one IIS instance. The IIS mapping is with a wild card or a specific host name. You can have several single sites with separate databases and code base on the same server. In that case, you have a separate admin interface for each site.

 >> Multi-site setup lets you have a single CMS site as a base (default site), and can create new sites in admin view that share the same root page, database and code base. The additional sites are either automatically mapped and require no additional configuration (if the base site is mapped to wild card), or they need manual configuration of host name.

When you work in a multi-site setup, you see all sites in the same interface, which makes it easy to work with them. One reason to run a multi-site setup with specific host name mapping (that is, a different IIS instance per CMS site) is that you can use different application pools, which means that if one site fails, the other sites continue to run.

Requirements

The following requirements must be met to manage websites in admin view:

>> Available licenses. A notification message informs you of the number of sites allowed by the license available for the installation. See License Information on the Admin tab for information. >> Unique URL. In admin view, each website must have a unique URL and start page in the content tree. Start pages cannot be nested.

>> Domain mapping must be configured in IIS.

>> For multi-site setup, the IIS must be configured to respond to any host name.

>> For single-site setup, each separate CMS site must have an IIS site configured.

Adding and updating a website from admin view

On the Websites tab, you can click a site to see detailed information about its settings. From here, you also can update the site information.

To add more sites to your installation, click **Add Site**. Add the following information when creating and updating site settings for your installation:

EXAMPLE: Default website with different host names, languages and redirection types

The following example shows a default website with several host names, languages and redirection types configured:

This example would lead to the following behavior:

» A request to http://redirected.se is redirected to http://examplesite.se using an HTTP 301 response.

» A request to http://www.examplesite.se is served the Swedish content.

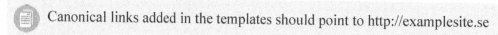 Canonical links added in the templates should point to http://examplesite.se

» A request to http://redirected.no/page/ is redirected to http://examplesite.no/page/ using an HTTP 301 response as this is the only Norwegian host that is not redirected.

» A request to http://redirected.com is redirected to http://examplesite.com using an HTTP 302 response as per the wild card specification.

EXAMPLE: Campaign website

The following example shows a campaign website:

Edit Website

Settings associated with the site.

Creating categories

 This topic is intended for administrators and developers with administration access rights in Episerver.

Episerver CMS, a category classifies content, such as for building filtering features for search and navigation. You create categories in admin view, and apply them to content in edit view.

 A category is a built-in property in Episerver CMS. You can apply categories to content, but you need to build the customized functionality for your website to display the resulting outcome, such as in a filtering. Also, do not confuse content categories with Commerce categories; see Creating a catalog entry.

Adding a category

Add a new top-level category as follows:

1. From admin view > Config tab, select Edit Categories.

2. Click Add. A new row is added to the table.

3. Specify a name in the Name field. This name is used in code when building category-based functionality.

4. Enter a name in the Display name field. This name is visible in edit view when a user selects categories. You also can language-encode this field.

5. Select Visible if you want this category to appear in the Select Categories dialog box in edit view.

6. Select Selectable if you want this category to be selectable in the Select Categories dialog box in edit view.

7. Use the up or down arrows if you want to move this category higher or lower in the list. This sequence determines the order in which categories appear in the Select Categories dialog box.

8. If you want to add a subcategory to a top-level, select the plus sign under Add. Add the subcategory in the same way as the top-level category.

Deleting or changing a category

Depending on how a category is used on your website, changing or deleting a category may cause things to stop working. Check with your developer before changing or deleting a category.

Editing frames and tabs

Frames

If you implemented frames on your website, you can use them for opening a link in a particular area of the window. This topic defined frames that are used by the system, and are accessible in the edit view as an option when an editor assigns a target frame for linking of content. Select the **Config** tab, and then **Edit Frames**.

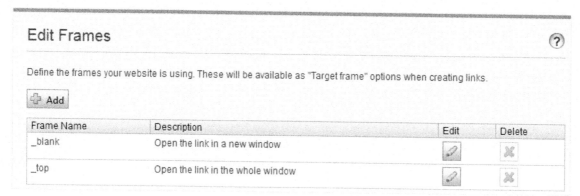

Edit Frames

Define the frames your website is using. These will be available as "Target frame" options when creating links.

Add

Frame Name	Description	Edit	Delete
_blank	Open the link in a new window		
_top	Open the link in the whole window		

Tabs

You can make properties appear on different tabs by selecting the **Edit Tabs** function. From here you can add, edit and delete tabs. You can also define the display order for tabs, and apply access levels.

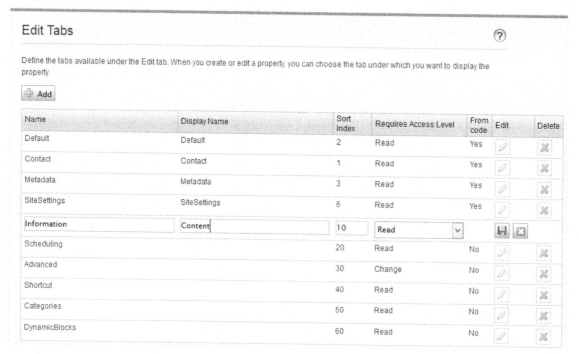

Edit Tabs

Define the tabs available under the Edit tab. When you create or edit a property, you can choose the tab under which you want to display the property.

Add

Name	Display Name	Sort Index	Requires Access Level	From code	Edit	Delete
Default	Default	2	Read	Yes		
Contact	Contact	1	Read	Yes		
Metadata	Metadata	3	Read	Yes		
SiteSettings	SiteSettings	6	Read	Yes		
Information	Content	10	Read			
Scheduling		20	Read	No		
Advanced		30	Change	No		
Shortcut		40	Read	No		
Categories		50	Read	No		
DynamicBlocks		60	Read	No		

Adding and editing a tab

1. On the Config tab, select Edit Tabs.

2. Click Add to create a new tab. Click the Edit icon to edit a tab.

3. In Tab, name the tab.

4. In Sort Index, specify the index amount for the tab. The lower the value, the further to the left the tab is placed.

5. In Requires Access Level, you can select which access level should apply for an editor to see the tab. It is linked to the editor's access level for the page.

6. Click Save.

Permissions for functions
Setting of access rights from Permissions for Functions

Set the access rights for the following functions, which are found in the admin view under **Config** > **Security** > **Permissions for functions**.

>> Detailed error messages for troubleshooting provides selected groups or users access to detailed error messages. In System Settings, you can activate a function that provides visitors with a form to fill in whenever a technical error occurs on the website. By changing the access rights here, you can specify who should receive these forms.

>> Allow the user to act as a web service user lets a user call one of the web services provided by Episerver. This function is used only for system integration purposes.

>> Allow users to move data/pages between page providers lets selected users or groups move pages between page providers. This is used for websites with a custom page provider integrated with another system. Because data is deleted in the source provider, you may want to limit access to this function.

Adding/Changing permissions to a function for a group or user

1. In admin view, go to Config > Security > Permissions for functions and select Edit for the function you want to modify. Existing groups or users with access appear.

2. Select Add Users/Groups if you want to give users or groups access to this function. The groups and persons in the system appear in the window that opens.

3. Double-click the name to add the group or user.

4. Select OK. The group or user appears in the list with its check box selected.

5. Click Save.

Deleting permissions to a function for a group or user

1. In admin view, go to Config > Security > Permissions for functions and select Edit for the function you want to modify. Existing groups or users with access appear in a list.

2. Clear the check box of the group or user for which you want to remove access.

3. Click Save.

For information about working with access rights, see Access rights.

CMS **Tool settings**

In admin view, under **Config** > **Tool Settings** you can perform miscellaneous functions for the integration and configuration of Episerver CMS.

>> Plug-in manager

>> Change log

>> Rebuild name for web addresses

>> Mirroring

>> Search configuration

Plug-in manager

Many functions in Episerver CMS are created as plug-ins, which you can manage from the **Plug-in Manager**. You can activate and deactivate selected parts. If your organization invested in additional plug-ins, you can find them in the Plug-in Manager also. After choosing a plug-in, choose which parts of it are accessible in the **Overview** tab.

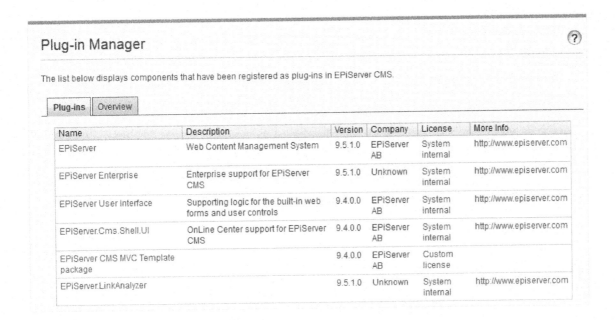

Change log

By default, all changes to pages, files and directories are logged in the **Change Log** system. You can filter the information in the Change Log, making it easier to find relevant information.

Changing the Change Log state

1.In admin view, select Config tab > Tool Settings > Change Log.

   2.Select a state:

>> Enabled. The Change Log starts automatically when the site starts and is available for read and write operations.

>> Disabled. The Change Log does not start when the site starts. Items written to the Change Log are ignored, but items may still be read from the Change Log.

>> Auto. The Change Log starts as soon as any dependencies (such as a Mirroring Job) are registered against it. If no dependencies exist, the system does not start. If already running, it stops.

Filtering the Change Log

1. On the View tab, filter and view change log items by entering one or several of the following values:

Field name	Description
Change date from	The query is run from the change log from this date.
Change date to	The query is run from the change log to this date.
Category	From the **Category** drop-down list, select: >> **Page** to run a query on pages only. >> **File** to run a query on files only. >> **Directory** to run a query on pages directories only. If you do not select an options from the drop-down list, changes are read from the **Change Log** when the query is run.
Action	You can filter the following actions in the Change Log: >> Check in >> Create >> Delete >> Delete language >> Move >> Publish >> Save >> Delete children
Changed by	To filter for a specific user, enter the Episerver CMS user name.
Maximum number of items per page	Limits the displayed number of items. Click the next and previous arrows to browse through the list of items.

2. Click Read to run the query. A list of matching change log items appear.

To remove all Change Log items that are more than one month old and without dependencies, use the Change log auto truncate scheduled job.

Website developers can customize and extend the Change Log. Consult your website developer for assistance.

Rebuild name for web addresses

Rebuild Name for Web Addresses changes addresses in the address field. When a visitor views a page on a website based on Episerver CMS, a path to the page appears in the address field. The address reflects the page's place in the website structure.

The names in the address field are created automatically from the name an editor specified for the page. If an editor changes the page name, the name in the address field does not change. You can manually change name in the address field by changing the **Name in URL** field on the **Settings** tab in edit view.

Some pages have no value in the field for names in web addresses, such as pages imported from other Episerver solutions. The **Rebuild Name for Web Addresses** function lets you create all web addresses for the website at the same time. You also can overwrite all existing addresses with new ones.

The **Rebuild Name for Web Addresses** function can affect links to the website. All internal links are updated automatically. However, if other websites link to a certain page, that link may be broken. The function also can affect visitors' browser favorites (bookmarks).

Creating a rebuilding name for web address

1. On the Config tab, select Rebuild Name for Web Addresses.

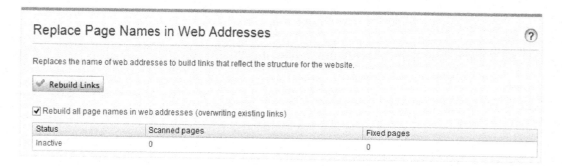

2. Select Rebuild all page names in web addresses (overwriting existing links).
3. Click Rebuild Links.

Mirroring

Mirroring duplicates content between websites. Episerver CMS can mirror selected parts or an entire website, and can run automatically or manually. This is useful if you wish to create sections in a test environment then publish all information at once to the public section.

Mirroring jobs are executed at time intervals that you set, as described in Scheduled jobs.

To enable mirroring, a mirroring application must be installed and running. The application handles data transfer between websites and is run separately to the Episerver CMS source and target sites. You can configure source and target websites to use separate mirroring applications. You also can install and configure a single mirroring application. See Episerver World for information about configuring and working with mirroring.

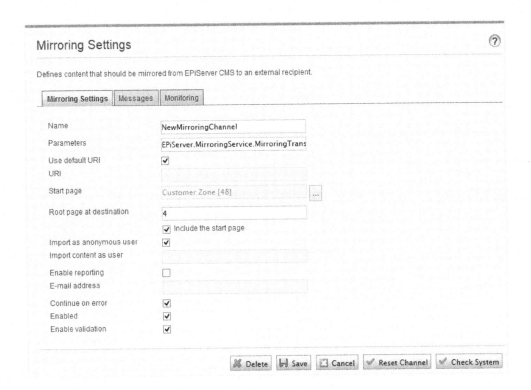

Creating a channel

To set up mirroring between two websites, create channels in the admin view. Channels define the mirroring jobs from one Episerver CMS to another, or even to an external recipient.

Search configuration

You can configure different **search providers** depending on the type of content on your website. A search provider searches across the website for pages, blocks, files, categories, forums or page types, or products on an e-commerce website. Search results are available when searching in the global Episerver menu in edit view.

Enabling search providers

You can decide which search providers you want to enable, and the order in which they appear in the search hit list. To access these settings, go to the admin view and select **Config** > **Search Configuration**.

>> Blocks. Allows for search in blocks.

>> Files. Allows for search in files.

>> Pages. Allows for search in pages.

>> Jump to. Makes it possible to jump directly from the search hit list to menu alternatives matching your search criteria.

» Products. Allows for search in products if you have Episerver Commerce installed. »
Find files, blocks and pages if you have Episerver Find installed.

You can drag and drop the search provider options to change the order between them. This controls the order in which the results are displayed in the hit list. Clearing a check box disables this search provider option.

CMS   Properties

Properties are central in Episerver CMS. Content type properties store and present data, such as page types and block types, and they are fields where editors enter information into a page type. A property can be the page name, an image placeholder, or the editor area, where you can add text. For example, the XHTML editor is a property of the type **XHTML String (>255)**, which results in an editorial area in the page type when used. For property content to be visible to visitors, link it to a content type with corresponding rendering.

Property types

You can define properties in code or from the administrative interface. For certain properties defined in code, you can create "non-breaking" changes in admin view to override these settings. If a property is defined in code and cannot be changed in admin view, information appears. If you make changes to a property defined in code, you can reset the changes to the values defined in code.

The following types of properties are used:

>> Built-in properties are set by the system and are automatically available for all pages and blocks regardless of type. For example, Page Name (name of the web page) and Page Start Publish (start publish date for the page).

>> User-defined properties are added to the page or block type definition in code or from the admin view. Typical examples are Heading and Main Body.

You can locate property settings under the **Config** tab in the **Property Configuration** section, and under the **Content Type** tab when you work with content types.

Editing and adding properties on content types

A common example for editing properties is to define toolbar buttons for the TinyMCE rich-text editor. You normally define properties in code but occasionally add them in admin view, because properties added from there are not rendered.

When you edit and add properties, the following tabs are available:

» Common Settings is where you edit a property's common settings.
» Custom Settings depend on the property data type being editing.

Editing a property

The upper **General** section of the **Common Settings** tab contains information about a selected property. If a property is **defined in code**, information about it is displayed, but you cannot change values such as property type and presentation control. You can change other settings, such as making a property mandatory or searchable. The lower **User Interface** part contains settings related to the property display in edit view.

See **Adding a Property** below for information about available settings for properties.

Adding a property

1. In admin view, from the Content Type tab, select a page type and click Add Property.
2. Fill in the fields.

Defining language-specific properties

When working with globalization, you define in every template which fields vary depending upon the language. To do this, set whether the property for that field should be "locked" or

"open" for globalization. This is done using the **Unique value per language** setting in admin view.

Properties that have unique value per language are editable in all enabled languages on the website, which is normally the case. You can edit only properties that do not have language-specific values in the language in which the page was created (the original page language). These properties are disables in edit view with an icon indicating the original language.

Imagine the property defining the sort order field is not set as a unique value per language (that is, the **Unique value per language** check box is cleared). When creating a new page, you can set sort order in the original page language. But if you create a version of the page in another language, the sort order field is not editable. As a result, the sort order is the same for all enabled languages. If you want to change the sort order for each language, select the **Unique value per language** check box.

If a property is changed to not having a unique value, all existing values for that property are deleted. So if the property for the editor area is changed to not have a unique language, all text entered in the editor area for all languages on the website are permanently deleted.

Setting a property to be language-specific

1. On the Content Type tab, select the page type that contains the property to be set.

2. Click the name of the property that you want to change.

3. Select the Unique value per language option.

4. Save your changes.

Organizing properties in content

You can alter the order in which properties are displayed to editors in a page or block type. You can also move properties between tabs in a page type. These changes are done for each page type in admin view.

Changing the order of properties

Follow these steps to change the order in which properties appear in the All Properties editing view. For example, you can display important properties at the top of the page.

5. Select the page or block type on the Page Type or Block Type tab.

1. Click an arrow to move a property, or drag and drop the property to a desired order.

Placing a property on a tab

1. Select the page type on the Content Type tab.
2. Click the name of the property that you wish to modify.
3. In the Tab drop-down list, select the tab on which the relevant property will be placed.
4. Click Save.

Configuring customized property settings

On the **Config** tab under **Property Configuration** > **Edit Custom Property Types**, you can configure custom property types.

Priorities and configuration

The property settings have the following priority:

1. A specific setting for a property defined in admin view. This can be either a custom settings for this property or pointing to a specific global setting.

2. A specific setting for a property defined for the model in code.

3. A global setting defined in admin marked as the "Default" settings for the property type.

4. A global setting defined in code.

Creating custom property types

The list of available custom property types done by a developer has the following columns:

  » Name. The name of the content type created by a developer, of which some are selectable in the editorial interface.

  » Base type. Shows the built-in types that can be extended by a developer.

  » Class name. Shows the full name of the class. The class is defined in the assembly.

  » Assembly name. Shows the class reference. A blank column indicates a built-in property.

You use two kinds of settings to change the editor's layout and buttons: **global settings** and **custom settings**. You configure the layout of the editor toolbar the same way regardless of the type of setting.

Changing the layout of the editor

1. Enter the required width and height of the editor (in pixels) in the Height and Width fields.

2. If needed, change the path to the cascading style sheet (CSS) that is used in the editor with the Content CSS Path field.

3. Drag and drop icons that you want available from the editor toolbar designer to and from the Inactive tools section.

4. Remove an icon by dragging it from the toolbar designer and dropping it in the Inactive tools section. The icon is automatically placed in the category to which it belongs.

5. Add an icon to the editor by dragging it from the Inactive tools section to the desired position in the toolbar designer.

6. Add and remove rows from the editor by clicking Add Row and Remove Last Row. The easiest way to clear all the icons from the toolbar designer and start from scratch is by clicking Clear Rows.

7. Configure the editor plug-ins and click Save.

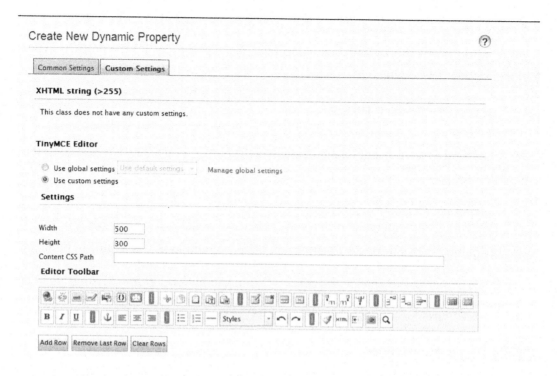

In the lower part of the **Create New/Edit Property** dialog, you can choose whether to use an advanced image or link dialog box. You also can turn on the word count in the editor, and specify if you want to use the Episerver CMS file selection dialog box. By default, the advanced image dialog box is used with the Episerver CMS file selection dialog box.

Enabling plug-ins

Some editor plug-ins are enabled always in a standard installation of Episerver CMS. These are configured in the Plug-in Manager on the **Config** tab in admin view.

» Change list buttons to advanced list buttons enables a drop-down list with advanced types for nested lists, such as square, lower alpha, and lower roman.

» Change image dialog to Advanced image dialog enables an advanced Add/Edit Image dialog box. Select this check box to configure advanced properties for your images.

» Word count enables word count functionality. Select this option to add a Words field to the bottom right of the editor. This displays the number of words included in the editor area.

Configuring global and custom settings

Global settings let you change the layout and the buttons in the rich-text editor toolbar for a property. You can use global settings on all page types as a default, or on only one page type. You can have as many global settings as you like on your website to create an editor suitable for all editors working with the website. You also can apply custom changes for a specific page type.

Configuring global settings for the XHTML String (>255) property

When you apply a global setting to all properties based on the **XHTML String (>255)** property type, all the editors on the website using a global setting are based on this, unless otherwise stated that a editor should be based on another global setting or a custom setting.

1. On the Page Type tab, select Edit Custom Property Types and click Add Setting.
2. Enter a descriptive name for the global setting.
3. Change the layout of the rich-text editor, configure the plug-ins, and click Save. The global setting appears in a list.

> Click **Set as Default** for one setting to use it for all the editor toolbars on the website, unless another setting is chosen for the property in a certain page type. If you do not configure a global setting as default, the properties use the standard toolbar set at installation.

Configuring global settings for a property on a page type

You can configure a global setting for a specific property on a specific page type.

1. Open the page type for which you want to change the global settings on the Page Type tab in admin view.
2. Click the property you want to configure and select the Custom Settings tab.
3. Select the Use global settings check box and select Use default settings if you want to use the default settings for the property.
4. Create a new global setting for the property by clicking Manage global settings.
5. Add a global setting by following the instructions on how to Configuring global settings for the XHTML String (>255) property.
6. Change the layout of the rich-text editor, configure the plug-ins, and click Save.
7. Change the global setting in the drop-down list and click Save.

Configuring custom settings
Use custom settings to change the layout and the buttons in the rich-text editor toolbar **for this property on this page type only**.

Configure a custom setting as follows:

1. On the Custom Settings tab, select the Use custom settings radio button.
2. Change the layout of the rich-text editor, configure the plug-ins and click Save.

» Deleting a global settings

To delete a global setting, open the setting and click **Delete**.

Content types

Content in Episerver can be page and block types, folders, or media files such as images and documents. Content also can be products in a product catalog in Episerver Commerce.

Content types and properties

Page and block types contain the properties where editors enter information, such as a link to an image on a web page.

For a content type, you define a set of properties that can hold information such as a link to an image or to a web page, or an editorial text. A typical website has a set of content types that match the identified functions needed on that website.

The content type is the base or blueprint from which you create one or many page or block instances.

To display content to visitors, the page or block type and its properties need to be mapped to corresponding rendering.

The content concept in Episerver is based on inheritance from a "generic" content type, which is then used to create specific content types, such as a page type or a media folder. Using this feature, developers can create custom content types easily when setting up new websites.

You can define page types either in code or from the admin view. On the other hand, you can define clock types only in code. For page types defined in code and for all block types, you can define "nonbreaking changes" of properties in admin view.

You are notified if you cannot change the settings defined in code in admin view. If you make changes to a page type defined in code, you can reset the changes to the original values defined in code.

Page types

 Be careful when you alter page type settings because changes may cause the website to stop working. Although you can create page types in admin view you should create them from code.

You cannot delete certain page types defined in code. These are typically page types upon which other page types are based, such as the standard or default page and the start page.

Creating a page type from admin view

Create a page type from the admin view as follows:

1. On the Page Type tab, select Create New Page Type.

2. Under the Information tab, enter the following information.

   Click Save, or Revert to Default if you want to restore your settings.

3. Go to the Default Values tab. You can set default values for some of the built-in properties in Episerver CMS from admin view. You also can specify default values for the properties from code, but these are not visible in the admin view.

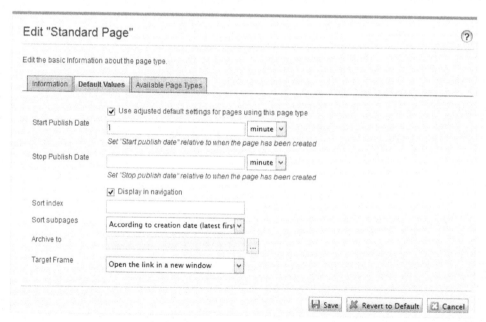

4. Change the following properties default values then click Save, (or Revert to Default if you want to restore your settings.)

5. settuings

6. Go to the Available Page Types tab. When creating new pages, you should limit the available page types in the page type list, to make it easier for editors to chose the correct page type. For example, for a News List parent page, you can specify that the only available page type is News Item.

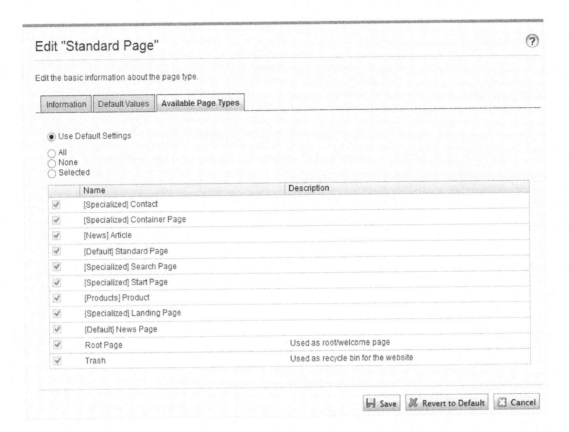

7. Define the desired page type by selecting any of the following options:

Option	Description
Use Default Settings	Select to revert to default settings as defined in code for this page type.
All	You can create pages based on all page types available in edit view.
None	You cannot create child pages for this page type.
Selected	Manually select the page types that should be available.

8. Click Save, or Revert to Default if you want to restore your settings.

Copying an existing page type

When you copy a page type, all of its properties are also copied. You can then edit the information for the page type and its properties. Copy a page type as follows:

1. Select Copy Page Type in the Page Type tab in admin view.
2. Select the page type that you want to copy from the drop-down list and click Copy. A window appears containing the exact same properties.
3. Edit the page type information to suit your requirements and click Save.

Block types

Block types are similar to page types, and you can modify some settings from admin view.

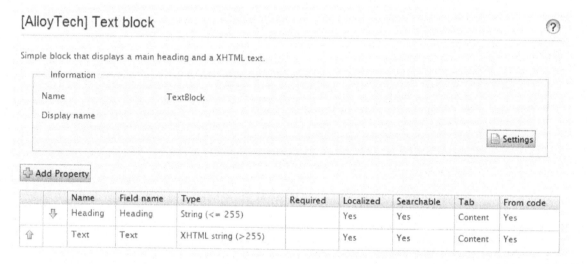

[AlloyTech] Text block

Simple block that displays a main heading and a XHTML text.

Information

Name   TextBlock

Display name

Settings

Add Property

		Name	Field name	Type	Required	Localized	Searchable	Tab	From code
	⬇	Heading	Heading	String (<= 255)		Yes	Yes	Content	Yes
⬆		Text	Text	XHTML string (>255)		Yes	Yes	Content	Yes

Editing a block type

Edit the settings for an existing block type as follows:

1. Select the block type on the Block Type tab and click Settings.

2. Change one or several of the settings as described for Creating a page type from admin view.

3. Click Save, (or Revert to Default if you want to restore your settings).

Other content types

By default, there are generic content types for other types of content such as folders and media files. Based on these, developers can create specific content types. You can have a specific folder or media file content type, where you can add properties such as *Copyright* or *Photographer*.

Visitor groups

Personalization in Episerver lets you target website content to selected visitor groups. The personalization feature is based on customized visitor groups that you create based on a set of **personalization criteria**. Episerver provides a set of basic criteria such as geographic location, number of visits, and referring search phrase. You can also develop customized criteria.

Managing visitor groups

Visitor groups are managed from the **Visitor Group** option in the global menu. When creating a new visitor group, you select one or more criteria and define appropriate settings. Those criteria are used to determine whether a user visiting the website is a part of that visitor group.

> You can make the editor a member of **VisitorGroupAdmins**. This group provides access only to the **Visitor Groups** option in the global menu.

Creating a visitor group

1. From the global menu, select CMS > Visitor Groups.
2. Click Create.

3. In the Criteria section, click to add criteria for the visitor group as follows:
    a. In the Match drop-down list, select the criteria to match All, Any or Point. What you select here affects all criteria for the visitor group. Using points is a way to set a value for what a desired action on the website is worth.

    b. Drag the criteria from the pane on the right and drop it into the Drop new criterion here area.

4. Make the settings for the criteria, see examples described in Examples of creating visitor groups.

5. In Name, name the visitor group you are creating. This name is displayed in the personalized content box when you select the content on a page.

6. In Notes, type a descriptive text about the visitor group you have created, for example, its purpose. This description is displayed as a tooltip when the editor is adding a visitor group to the content on a page.

7. In Security Role, select the check box if you want this visitor group to be available when setting access rights for pages and files in admin view. Note that visitor groups only have read access.

8. In Statistics, keep the check box selected to enable statistics for the visitor group (this check box is selected by default).

9. Click Save.

Editing a visitor group

1. From the global menu, select **CMS > Visitor Groups**.

2. Click Edit for the visitor group you want to change.

3. Add a new criterion for the visitor group by drag-and-drop, change the value for an existing

Copying a visitor group

1. From the global menu, select CMS > Visitor Groups.

2. Click Copy for the visitor group you want to copy. The new copy has the same name as the original but with the extension "- Copy".

3. Rename and change criteria for the new visitor group you have copied.

Deleting a visitor group

1. From the global menu, select CMS > Visitor Groups.

2. Click Delete for the visitor group you want to delete.

3. Confirm the deletion.

Viewing and clearing statistics

The visitor group statistics appear as a gadget on the dashboard. Clear the statistics from the database as follows:

Examples of creating visitor groups

EXAMPLE: Site criteria and points

By using **Points**, you can set a value for how much an action is worth, for example, a visited campaign page. In this example, a visitor who visits the page on a certain date matches the visitor group criteria.

1. In the Match drop-down list, select the criteria to match Point.

   Drag and drop the Visited Page criterion, and select page. Use drag-and-drop of the criteria again to add several pages.

2. Drag and drop the Number of Visits criterion, and select More than > 1 > Since [date]. (To create a visitor group for visitors who have never visited the page, select Less than > 1 > Since [date].)

3. Enter the number of points each criterion is worth, and select whether or not the criterion is required.

4. Select Threshold for the criteria you added in your visitor group. For example, the visitor must fulfill 1 of 3 criteria to be included in the visitor group.

5. Click Save.

EXAMPLE: Geographic location

You can direct content to visitors from a specific country and specific days. For example, people from Sweden visiting your website on weekdays. You can show these visitors a clickable banner to sign up for a conference.

1. Drag and drop the Geographic Location criteria, and select Continent, Country and/or Region. Use drag-and-drop of the criteria again to add several countries.

2. Drag and drop the Time of Day criteria, and select [weekday]. You can also select the personalization to start and end at a specific time.

3. Click Save.

EXAMPLE: Geographic coordinate

You can direct your content to visitors from a specific part of a city, for example, "People from Upplandsgatan, Stockholm".

1. Drag and drop the Geographic Coordinate criteria, and click Select Location.

2. Click Select location to display a map that you click to set a location. You can zoom in the map for more precise locations.

3. Select the Radius [number of kilometers or miles].

4. Click Save.

EXAMPLE: Referrer

The HTTP Referrer is based on pages, such as those used in a campaign. For example, you can target content to visitors who

>> search for "episerver and cms" on Google.com >> from the search result page, click the Episerver landing page link

>> Drag and drop the Referrer criteria, and select URL > Equals > the URL of the search engine result page, for example, http://www.google.se/#hl=sv&source=hp&biw=1338&bih=790&q=episerver+cms.

>> You also can add the Geographic Location to select a country.
>> Click Save.

EXAMPLE: Form and form values

You can base a visitor group on whether the visitor has (or has not) submitted a particular form or form value. For example, you can target content to visitors who submit a Job Application form:

You can target content to visitors that give you a low rating on a Satisfaction Survey.

Additional visitor group criteria

You can extend the built-in visitor group criteria as follows:

Commerce Episerver Commerce criteria

Visitor group criteria specific for e-commerce, such as customer properties, markets, and order frequency criteria. See Personalization for Commerce.

`Addons` **Episerver Visitor Group Criteria Pack**

» Display Channel matches the visitor's current display channel when visiting the website, such as distinguishing between web and mobile visitors.

» IP Range matches the IP range either equal to, below, or above a defined IP number the visitor used when visiting the website.

» OS & Browser matches the operating system and browser the visitor used when visiting the website.

» Role matches the access roles the visitor had when visiting the website. You can either include roles by using the In role condition, or exclude roles by Not in role.

`Addons` **Episerver Marketing Automation**

Visitor group criteria specifically designed for marketing automation

Episerver CMS
Developer
Guide

Introduction

In this course, We need to focus on the core functionality of the Episerver CMS, it's various parts, from building the models, to controllers and down to the layout and views. We will build a site from scratch following the look and feel of the Alloy Samples Site.

I Understanding EPiServer CMS

Imagine that we've been asked to build an iPhone app. Also imagine that we've never used an iOS device, or perhaps not even a smartphone or tablet of any kind. We could of course fire up Xcode, the IDE used when building iPhone apps, and start coding away. But without knowing anything about the context in which our app runs odds are we wouldn't get very far. Even if we did manage to build a working app our users probably wouldn't find it very easy or enjoyable to use - we have no idea about how iPhone users are used to interacting with apps.

The same goes for EPiServer CMS development. In order to build an EPiServer site we need to know how our users, the editors, interact with the system. In this part of the book we'll introduce the core concepts of EPiServer CMS, and many other CMS like it. Then we'll walk through the steps required to install the CMS and create a sample site. Equally important we'll also look at what the installation does. Finally, we'll use the sample site that we learn to set up to learn the most important aspects of using the CMS to create and edit content on a website.

1 From static site to CMS

EPiServer CMS is a complicated product that has been developed for over a decade. As you can imagine, it's very rich in terms of features, some more useful than others. This creates a risk when writing, and reading, a book about EPiServer development; we may lose focus of the core concepts. If that were to happen we'd gain a superficial understanding of many fringe features without actually understanding the essence of EPiServer CMS.

Therefore, to allow us to focus on the essence of EPiServer CMS and EPiServer development this chapter, the first chapter in this book about EPiServer CMS development, actually isn't about EPiServer CMS.

Instead it's a story about a site that doesn't exist.

We'll follow this fictive site from its infancy, being a single static HTML page, to a dynamic, database driven website where non-technical users can create and edit its content. In the process we'll face the same problems that the early developers of EPiServer faced and we'll, for the most part, implement the same solutions.

While we'll look at code in this chapter, as with the site, this code isn't real. Instead it's pseudo code intended to illustrate solutions and concepts. So, don't try to compile it :-)

In the beginning there was HTML

Imagine that we've just been hired to work as web developers by a company that imports and sells fruit The company isn't present on the web and it's our job to change that.

As a first step we create a single page website with some information about the company along with contact information

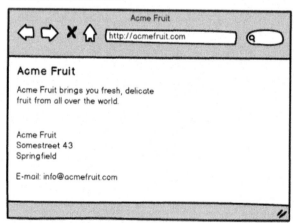

Simple image and text only website(change picture)

All that's needed for this simple site is a static HTML file and the logotype. In order to publish it we upload it to a web server using FTP.

The markup in the HTML file, index.html, looks something like this:

```
<html>
  <head>
    <title>Acme Fruit</title>
  </head>
<body>
  <img src="logo.png" />
  <p>
Acme Fruit brings you fresh, delicate
fruit from all over the world.
</p>
<p>
Acme Fruit<br />
Somestreet 43<br />
Springfield
</p>
<p>E-mail: info@acmefruit.com</p>
  </body>
</html>
```

Next we want to add information about the company's products. We do that by adding a new static HTML page similar to the first one.

As we've now left the single page site stage we also decide to move the contact information from the start page onto its own separate page.

Three static HTML pages

Of course having multiple pages on our site we need a way for visitors to navigate between them. To provide that we add a navigation bar to each of the pages. The navigation shows what page the visitor is currently viewing and contains links to the other pages on the site.

The products page with top navigation(change image).

The markup in the HTML file, index.html, looks something like this:

```
<html>
  <head>
    <title>Acme Fruit</title>
  </head>
<body>
  <img src="logo.png" />
  <p>
Acme Fruit brings you fresh, delicate
fruit from all over the world.
</p>
<p>
Acme Fruit<br />
Somestreet 43<br />
Springfield
</p>
<p>E-mail: info@acmefruit.com</p>
  </body>
</html>
```

Next we want to add information about the company's products. We do that by adding a new static HTML page similar to the first one.

As we've now left the single page site stage we also decide to move the contact information from the start page onto its own separate page.

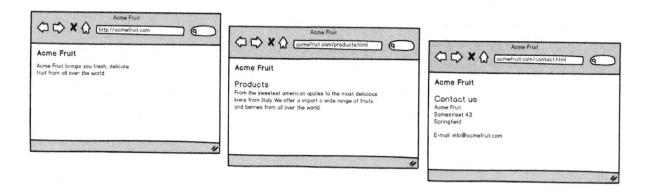

Three static HTML pages

Of course having multiple pages on our site we need a way for visitors to navigate between them. To provide that we add a navigation bar to each of the pages. The navigation shows what page the visitor is currently viewing and contains links to the other pages on the site.

The products page with top navigation(change image)

The HTML for the navigation bar looks something like this on the Products page:

```
<ul>
<li><a href="/">Home</a></li>
<li>Products</li>
<li><a href="/contact.html">Contact</li>>
</ul>
```

Apart from the fact that the page that the visitor is currently viewing isn't linked to in order to indicate

that it's the currently viewed page the code for the navigation is identical on the three pages.

Componentization and server side processing

Next we decide to add a fourth page with information about the company. Following the workflow that we've used so far, using static HTML pages, the easiest way of adding the new page would be to copy one of the existing pages and change the copied page's content. While that's easy we would also have to add a new link to the top navigation. As the code for the navigation exists on all four pages and we want the link to the new page to exist on all of them we'll have to first add it to the navigation on one page and then copy and paste it into the other three pages. This is tedious and error prone work and as we envision

that the site will continue to grow it's clear that our current solution for the navigation won't work, at least not efficiently.

The problem with the navigation is twofold. It is made up of nearly identical *duplicated code*, meaning that in order to make a single change to its functionality we have to make changes to code in multiple places. Second, it's made up of *static code*. While it differs slightly depending on what page it's displayed on as the name of the currently viewed page is treated specially, this logic is handled by the fact that the code for the navigation is duplicated on each page with minor variations. It's of course also static in the sense that there's no logic or data source that determines what links it should show - the links are hard

coded.

If it hadn't been for the slight difference in the code for the navigation that highlights the currently viewed page there would have been a simple solution to the code duplication problem, we could have used a technology called Server Side Includes (SSI). SSI is supported by most popular web servers. When using SSI the web server processes HTML files before serving them to the client.

During this processing it looks for a special tag in the HTML code that specifies a path to another HTML file. Upon finding such a tag the web server replaces it with the content in the other referenced file.

SSI is a simple solution for extracting and reusing chunks of HTML on multiple HTML pages. As such it allows for basic componentization, often a prerequisite for maintainable and modular software applications. However, while SSI enables reuse of static HTML it doesn't allow for any other type of server side processing. Therefore we can't use SSI to make our navigation easier to change as we also need to implement logic to handle what link is highlighted depending on what page the navigation is being displayed on.

To address that we need more advanced server side processing functionality.

We need a *web application framework*. A web application framework typically acts as a plug-in to the web server and provides an alternate pipeline for handling requests, often for files with a specific file extension.

When the web server receives a request for a file of a type handled by the application framework it deviates from its standard

procedure of locating the file on the harddrive and instead hands over processing to the application framework.

After being handed over control of the processing of a request, along with information about it such as what file or path was requested, from the web server the framework typically proceeds to execute source code in the requested file.

The output from the code in the requested file is written to the response sent

to the client from which the request originated.

While a web server can handle serving a wide range of static files a web framework only handles a limited set of files.

On the other hand, as opposed to a web server, an application framework cares about what's in the files it does serve, typically executing application specific source code in the files. Beyond executing source code a web framework also provides an environment in which the code is executed. By means of this environment the application specific source code can interact with the incoming request, the outgoing response as well as a host of other resources and functionality, such as performing operations on files on the file system, accessing databases and utilizing caching.

What programming language the application specific source code is written in differs from framework to framework.

So does the environment, operating system, web server etc in which the code is executed. Some frameworks provide a lot of functionality out-of-the-box, often coupled with a number of abstractions and reusable components. Other frameworks takes a minimalistic approach focusing on flexibility for developers rather than rapid development of common functionality.

There exists a wealth of web frameworks for us to choose from when developing websites. Some are commercial and must be licensed. Some are free (and often open source) and can be used without restrictions and some are provided as part of, or extensions to, other software such as web servers or operating systems.

Some of the most popular web application frameworks include PHP, which is both a programming language and a web framework, Ruby on Rails in which functionality is developed in the programming language Ruby and JavaServer Pages (JSP) in which development is done in Java. When a website is to be hosted on servers running Microsoft Windows the most commonly used framework is Microsoft's own ASP.NET.

ASP.NET allows for web site development in any programming language that runs on the .NET execution engine, the Common Language Runtime (CLR). In practice this usually means C# or VB.NET although there are other alternative CLR languages, including F# and Boo. When building sites the ASP.NET framework offers two "flavors", Web Forms and MVC.

Any self respecting web framework can be used to solve our problem with the navigation as they provide functionality to add logic to the handling of requests as well as some sort of ability to break pages, or From static site to CMS 6

"views", up into components for reuse. However, as the company that we work for prefers Microsoft technology we decide to use ASP.NET.

Assuming we've decided on a framework, we create a new project and convert all of our existing HTML pages into files handled by that framework. With ASP.NET that means either .

ASPX files when using Web Forms or controllers plus views when using MVC. We then extract the navigation from one of the pages into a reusable component, such as an User Control (.ASCX file) when using Web Forms or a partial

action or HTML helper method when using MVC.

Having created the partial component with the static HTML code from one of the pages we replace the navigation in each of the pages with it. We've now fixed the code duplication but in the process we've introduced a bug.

No matter which page is viewed the navigation will highlight the one from which

we copied the code. To correct that we need to add some logic to the component so that it's displayed differently depending on what page it's rendered on.

As a first step we create a small class that represents an individual link in the navigation. For that purpose it needs three properties, the link's text, the URL to link to and a boolean indicating whether it should be displayed as the currently viewed page.

```csharp
public class MenuLink
{
public string Text { get; set; }
public string Url { get; set; }
public bool Selected { get; set;
}
```

Then, in the component, such as in the user control's code behind file or in a controller action, we create a method that returns a number of instance of this class, one for each page on the site.

```csharp
public IEnumerable<MenuLink> GetMenuLinks()
{
    var links = new List<MenuLink>
    {
      new MenuLink
      {
         Text = "Home",
          Url = "/"
      },
      new MenuLink
    {
    Text = "Products",
    Url = "/products"
    },
  new MenuLink
{
    Text = "About us",
    Url = "/about"
},
  new MenuLink
{
Text = "Contact us",
```

```
   Url = "/contact"
  }
};
return links;
}
```

While we haven't yet implemented any functionality to set the Selected property to true for the currently viewed page's link we can now replace the hard coded HTML in the component with a loop that iterates over the objects returned by the method and render a HTML link for each of them.

Pseudo code for that may look like this:

```
<ul>
```

```
foreach(var link in GetMenuLinks())
{
<li>
if(link.Selected)
{
<%= link.Text %>
} else
{
<a href="<%= link.Url %>"><%=link.Text%></a>
}
</li>
}
</ul>
```

Now the menu is generated from the values returned by the GetMenuLinks method and the component renders each page as linked or not depending on the Selected property. To finalize the navigation we need to set the Selected property to true for the currently viewed page.

If a link's URL matches the requested URL we mark it as selected.

```
...
foreach(var link in links)
{
link.Selected = Request.Url.EndsWith(link.Url);
}
return links;
```

Layouts

The code for the navigation is now centralized to a single place, making it far easier to update when adding new pages to the site.

We can continue to use this concept of extracting common functionality to components such as partial views, HTML helpers or user controls. We may for instance want to move the From static site to CMS 9 code for the logotype into a reusable component making future changes to it easier.

There is however one type of code duplication for which this type of componentization isn't suitable; layout.

We'd now like to add a footer to the site with some copyright information and possibly some links. We could of course add the HTML code needed for the footer to each individual page but that would create a similar problem as the one we just solved for the navigation. Therefore we can instead create a reusable component that renders the footer.

Still, after creating the footer component we would need to need to go through each of the pages on the site and add the component. This highlights the fact that the site's layout, it's graphical framework if you will, is duplicated and hard to maintain. That is, while we can reuse individual code blocks by breaking them out into components, adding or moving such components that should be used across the entire site demands that we modify each individual page, even though all of the pages share the same graphical layout.

To address this problem we need to separate the shared layout from the parts of the pages that is actually different on each page, typically the main editorial content. Most web frameworks or view engines offer

some sort of functionality to accomplish this, although both the implementation and name for it may be different. In ASP.NET with Web Forms this functionality is called master pages.

In ASP.NET MVC it depends on what view engine is used. In version four of the MVC framework the default view engine is named Razor and its solution to this problem is called simply layouts.

Both master pages and layouts in Razor acts as a top level framework for individual pages, or for more specialized master pages or layouts. Master pages allow us to define placeholders called content placeholders.

A content placeholder can be empty or contain default content, including other content placeholders.

A page, or another master page, can specify that it uses a master page. When doing so we're not allowed to put HTML code anywhere we'd like in the page but can instead override one or several of the master page's content placeholders. When the page is rendered the framework will see that the page uses a master page and will render the master page with any content added, or "filled in", by the requested page.

Razor layouts are similar but instead of them being able to define multiple areas where a view, or another layout, can provide content they must somewhere within the code invoke a method named RenderBody.

When an individual view is rendered the framework sees that it uses a layout and renders that with the content of the view in place of the RenderBody method. In addition to this layouts can define additional extension areas for views called "areas". A view can then populate a given area by declaring that a piece

of the view should be rendered inside a named area.

This type of functionality is great for defining top level HTML code which defines the common layout for all pages, as well as common elements, such as the logotype and the footer in our case. Naturally, the first step in using a layout file of some sort is to create it. We can then extract the HTML code that defines the layout for the site from one of the pages. In the process we also add the footer.

```
<html>
<head>
<title>Acme Fruit</title>
</head>
<body>
<img src="logo.png" />
<% TopMenuComponent %>
<div class="footer">Copyright Acme Fruit</div>
</body>
</html>
```

Next we define a place holder in the layout where content from pages that use the layout will be inserted.

So far our site is so simple that there's only one part that differs between each page; the content between

the top menu and the footer. Therefore we insert a placeholder there in the layout.

```
...
<% TopMenuComponent %>
<% Placeholder_for_content_from_pages %>
<div class="footer">Copyright Acme Fruit</div>
...
```

We can now start using our layout by adding a directive to each page configuring them to use the layout.

We also remove the common HTML code and leave only the code and content that is specific to each page.

```
<% UseLayoutDirective %>
<h1>Welcome to Acme Fruit</h1>
<p>...</p>
```

Following the shift to using a layout we now have a more flexible and maintainable site compared to when we had four static HTML files all defining the same "framework" elements.

When we add a new page we need to make sure that it uses the layout and add link to it in the top menu component but can then focus on the content in the page.

When we want to change the design of the site we simply need to modify the layout file.

Storing page content in a database

The site in its current state is fairly easy to work with given that we know HTML, have a basic understanding of ASP.NET and know how to transfer files using FTP. While this knowledge is common among developers it's typically not developers that publish and update website content in an enterprise.

The company that we work for is no exception. While everyone in the organization is excited about having the website we soon start getting more and more requests to update its content.

This gets in the way of further development and those requesting the content changes feel frustrated as they have to wait for us to implement the changes.

Currently the content in each page is stored in a separate file.

If we moved the content into a database and added an authentication mechanism to the site we could create an interface through which non-developers could work with the content.

As a first step towards that we create a database and add a table where each row will represent a page.

The table will need to have an identity column, a column for the pages name in as displayed in the top menu, we'll call that "Name", and a column for the content. Having created the table and populated it with the data from each page it looks like this:

id	name	content
1	Home	<h1>Acme Fruits…
2	Products	<h1>…
3	Contact	<h1>…
4	About	<h1>…

As we've now moved all of the page specific content into the database we can remove all but one of our existing pages (ASPX files when using Web Forms or Controllers and or Views when using MVC).

We'll keep the one that we used as the start page and use that as a template for all of our pages, rendering the content stored in the database.

Of course, for that to work we'll need to know which page's content we should render in the template.

As each of the pages have a unique integer identifier, from the id column in the database, we'll use a query string parameter named "id" for handling that. When the template is called upon to handle a request we make it check whether the request contains such a query string parameter.

If it does, we'll fetch that row from the database and render the content in it.

If it doesn't, or the query string parameter is invalid, we make the template default to the ID of the start page (1 in the table above).

In (pseudo code) practice the code for this consists of two methods. The first returns the ID of the page that should be rendered, the current page:

```
public int GetCurrentPageId()
{
var id = 1;
if(Request["id"] == null)
{
return id;
}
int.TryParse(Request["id"], out id);
return id;
}
```

The second method uses the above method to fetch and return the content for the current page:

```
public string GetCurrentPageContent()
{
var sql = "select content from page "
+ "where id = " + GetCurrentPageId();
//Execute the SQL query against the database
//and return the result.
```

THE MISSING MANUAL

}

Finally we modify our remaining page, or rather our template, to display the content.

```
<% UseLayoutDirective %>
<%= GetCurrentPageContent() %>
<!-- The below static HTML is replaced by the line above.
<h1>Welcome to Acme Fruit</h1>
<p>...</p>
```

-->

Our site now stores and fetches its content from a database. A request to a URL such as http://acmefruit.com/default.aspx?id=or http://acmefruit.com/home/?id=1 displays the start page. A request to /default.aspx?id=2 or /home/?id=2

displays the Products page and so on.

As a result the top menu is broken as the URLs for all pages except the start page has changed. The top

menu also still uses hard coded texts for the links. We fix both issues by modifying the top menu so that

it fetches ID and name for all pages in the database and renders links to them.

```
public IEnumerable<MenuLink> GetMenuLinks()
{
string templatePath = //Relative URL for the path,
//such as ~/default.aspx or /home/.
var sql = "select id, name from page";
//Execute the SQL query against the database and
//assign the result to a DataReader represented by
//a variable named reader.
var links = new List<MenuLink>();
while(reader.read())
{
var pageId = reader.readInt(0);
var pageName = reader.readString(1);
var url = templatePath + "?id=" + pageId;
var selected = pageId == GetCurrentPageId();
var link = new MenuLink
{
Text = pageName,
Url = url,
Selected = selected
};
}
return links;
}
```

A user interface for working with pages

With the top menu updated both the content and the navigation of the site is based on what's in the database.

Adding, updating or deleting pages no longer require changes to the code. It does however require access to the database. Worse, it requires knowledge about databases. We're halfway to making the site manageable for non-technical users but we lack an interface for editing content.

To address that we add a new handler (ASPX or MVC controller + view). We don't make this new handler use the layout as this page won't be for public display on the site and therefore doesn't need elements such as the top menu and the footer.

Instead we add functionality to the handler to generate a list of links for all pages in the database.

While the code will be similar to the top menu this list is vertical instead of horizontal. Also, it doesn't link to the template used for public rendering of pages but instead to yet another handler that we'll create that we'll call "edit page handler".

That handler will still need to know what page has been selected though,

so the URL will be made up of the URL for the handler and an "id" query string parameter.

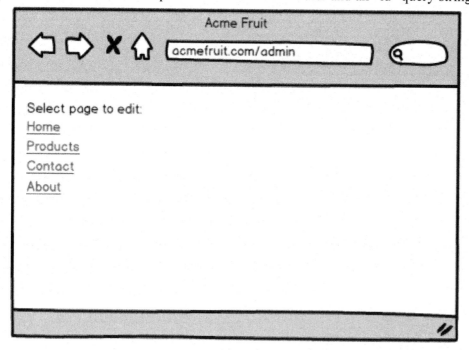

Next we add the edit page handler. It will render a textbox for the name, a textarea for the content and two buttons, one for saving and one for canceling editing without saving.

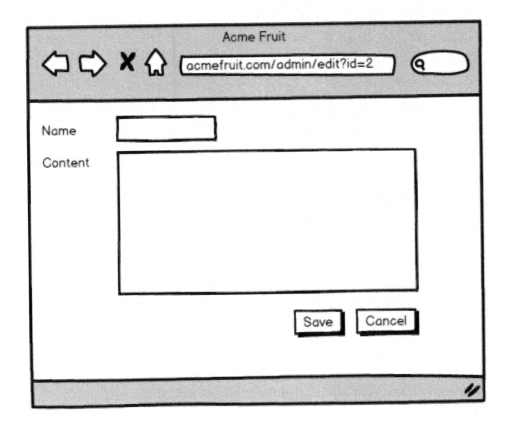

To enable users to edit a page we first need to fetch the content. We do so by retrieving the page's id from the query string and then fetch all cells from the database row with that ID.

Once fetched from the database we assign the value of the name cell to the textbox and the content cell's value to the textarea.

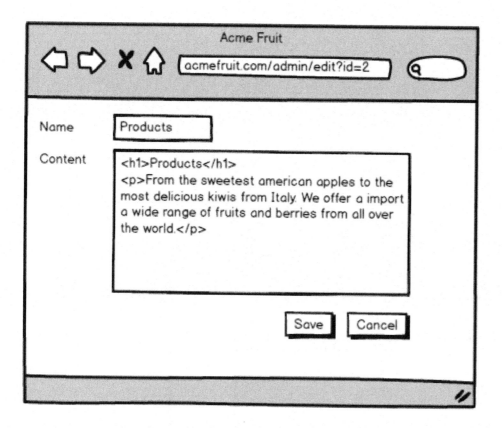

While users are now able to look at the values stored in the database as well as changing them in their browser they aren't yet able to save the changes. Therefore we proceed by implementing the buttons.

If a user clicks the Cancel button we simple redirect them back to the list of pages. However, if the user clicks the Save button we read the values from the Name textbox and the Content textarea and update the page's row in the database, locating it by ID using the "id" query string parameter. After successful updating it we display a message confirming that the page was indeed successfully updated.

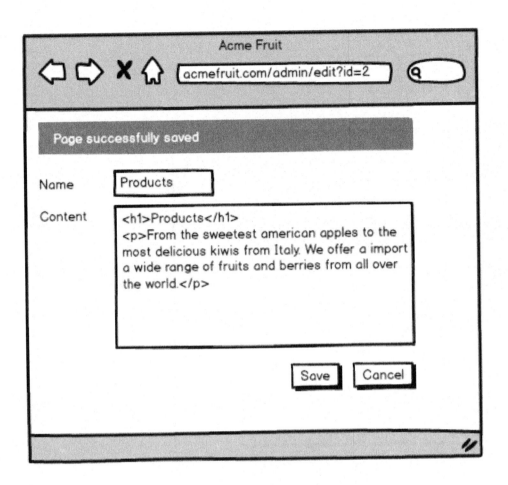

Great! Users are now able to modify existing pages through a graphical interface. However, we have one rather serious bug; if a user changes the name of a page to a blank string it won't be possible to navigate to it on the public site as there won't be anything to click on in the top menu. Worse, there won't be anything to click on in the list of pages for editing either, making it hard to revert the change.

Therefore we add some validation. Prior to updating the page's row in the database we check that the name isn't empty. If it is we don't execute the database update and instead show a

an error message.

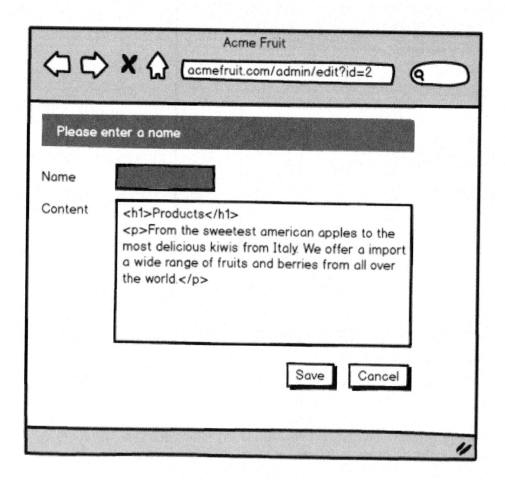

Our users are excited that they can now change the content on existing pages as long as they have a basic understanding of HTML. However, they would really like to be able to create new pages as well.

To address that we add a button for creating new pages.

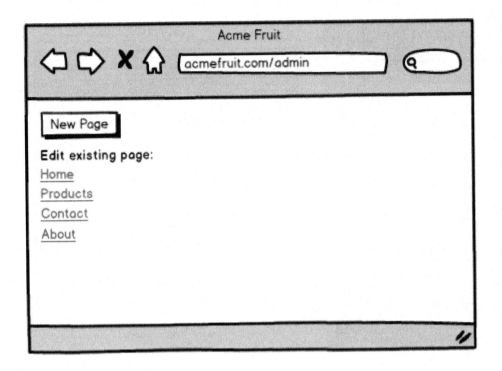

When a user clicks the button we redirect them to the edit page handler but without an "id" query string parameter. In the code for the edit page handler we implement a check whether an existing page has been selected. If not we simply skip populating the textbox and textarea upon loading.

Also, when a user presses the Save button we perform the same check and if we deduce that it's not an existing page that's being edited we use a slightly different SQL statement, using INSERT instead of UPDATE.

A minor but important detail is also that we after having created the new page don't redirect the user back to the same URL but instead add the newly created page's ID to the URL in the form of the "id" query string parameter.

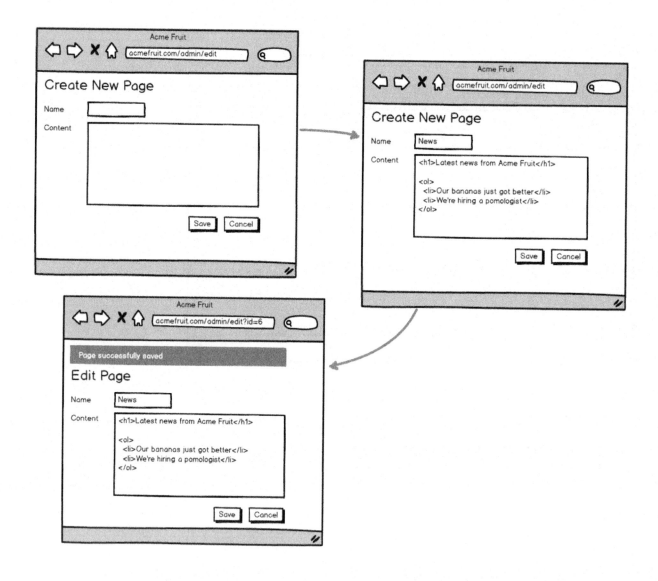

Pseudo code for the functionality for saving a page in the edit page handler looks something like this:

public void Save()

```csharp
{
if(string.IsNullOrEmpty(input["name"]))
{
//Show validation error
return;
}
if(Request["id"] == null)
```

```
{
//Create INSERT SQL statement and execute.
//Retrieve returned ID from the database.
//Redirect to the same handler with ?id={returned_id}
//added to the URL.
}
else
{
//Create and execute UPDATE SQL statement based on input
//and ID from query string.
//Redirect to the same URL.
}
}
```

Of course, after enabling users to create new pages we don't have to wait long for them to ask for functionality to remove pages. So, we add a delete button to the edit page dialog that is only shown when editing an existing page.

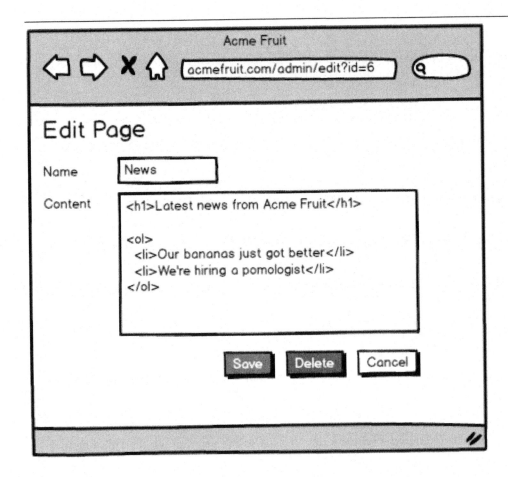

Acme Fruit

acmefruit.com/admin/edit?id=6

Edit Page

Name   News

Content
```
<h1>Latest news from Acme Fruit</h1>

<ol>
  <li>Our bananas just got better</li>
  <li>We're hiring a pomologist</li>
</ol>
```

Save   Delete   Cancel

When a user presses the Delete button we execute a DELETE SQL statement against the database with the page's ID from the query string in the WHERE clause. After that we redirect the user back to the list of pages.

We now have a simple but effective user interface for managing content on a small site, such as ours.

Nontechnical users can create, update and remove pages by simply redirecting their browsers to a special part of the site.

While everyone is happy that the site can be updated without having to upload files using FTP or by manually updating a database our users still complain about having to write HTML when editing the content in pages though.

To address that we modify the edit page dialog to use a WYSIWYG (What-You-See-Is-What-You-Get) editor.

Such an editor takes HTML content in an existing input element and let users modify it without having to write HTML code.

Instead they see only the rendered content and can work with it using various

tools in the editor, such as for selecting headers or font styles, that modify the underlying HTML for them.

When the underlying HTML is updated the editor writes it back to the original input field, meaning that we as developers still deal with the "raw" HTML.

While we could in theory create such an editor ourselves using JavaScript or some other client side technology such as Flash or Silverlight that would be borderline professional misconduct considering the rich flora of existing editors. After examining the market for WYSIWYG HTML editors we decide to use TinyMCE, a free JavaScript based editor with a rich ecosystem of plug-ins.

Adding it to our edit page dialog doesn't require us to change any of the server side logic. We do however have to add the JavaScript and CSS files needed to use TinyMCE as well as some custom JavaScript that replaces the existing textarea with the editor.

After doing so users no longer have to manually change the HTML markup but can instead edit it in a way resembling how they would use a word processor such as

Microsoft Word.

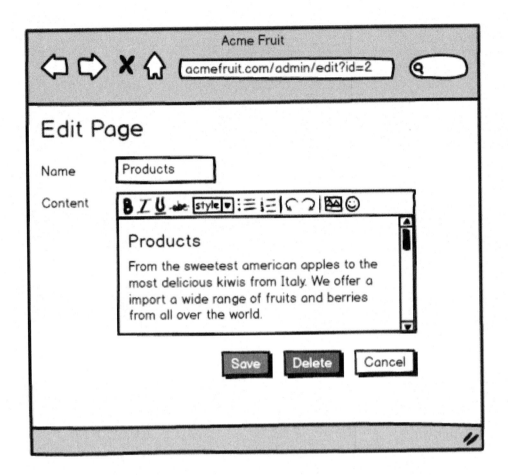

After saving a page the content, when viewed on the site, may not look exactly as it did in the editor as it's rendered in a different context using different CSS styles. However, our users are still happy as they can now create paragraphs by simply pressing the enter key and create headlines, bulleted lists and

other types of formatting without having to write HTML code. The users are able to focus on working with content rather than dealing with the intricacies of the underlying HTML used for structuring and formatting it.

As developers we're also happy as the WYSIWYG editor does its magic on the client side and when the

content is submitted we still get the raw HTML as a string which we can save to the database. Just as before.

Site development

Having covered what EPiServer CMS is and how it's used to edit content in part one the time has (finally) come to learn how to develop websites with EPiServer. In this part of the book we'll focus on the core concepts of EPiServer CMS development and create a simple but fully functional site.

After a look at core classes and concepts in chapter four and some essential tools in chapter five we get to hands-on development in the subsequent chapters, starting from scratch by creating a new, empty EPiServer site project and building and gradually turning it into a full blown site with page types, templates and dynamic navigation components.

Core concepts

Learning software development technologies is often a case of the chicken or the egg dilemma. If we start with how to build something we need to take the code at face value without understanding the underlying concepts. If we on the other hand start with the underlying concepts things tend to become abstract and

frustrating.

To address this problem, without having found a magical solution for it, we'll start with a minimum of core concepts and classes in this chapter, and then elaborate on them as we get hands-on with actual development. So, before we delve into hands-on development let's cover some theory.

PageData

The PageData class located in the EPiServer.Core namespace is used to represent pages in EPiServer CMS. That is, each page shown on in the page tree in edit mode is an instance of the PageData class.

In general a PageData object is synonymous to a web page on the site.

The PageData class has a lot of members. Some of the most significant are listed below:

PageName : string

The page's name that is displayed in the page tree and in other places in edit mode. The page name is often also used as the page's heading in templates.

Item : Object Allows access to a page's property values using indexer syntax. For instance, to retrieve a page's name without using the PageName property the following can be used:

PageData page; *//We assume page has been initialized.*

var name = page["PageName"] **as string**;

PageTypeID : int

The ID of the page's page type. See the section about page types below.

PageLink : PageReference

The PageReference object that is the unique identifier of this page. Think of this property as the ID of the page. See the section about ContentReference and PageReference below.

VisibleInMenu : bool

A built in property for all pages that is intended to be used when rendering navigation components on the site. If the property returns false the page should not be displayed in navigations/menus. The page is however still fully accessible to public visitors. The property corresponds to the "Display in navigation" checkbox in the top area when editing a page in forms editing mode.

The checkbox in forms editing that corresponds to the VisibleInMenu property.

Page types and content types

While all pages are represented as PageData objects in the CMS, pages can have different characteristics in the form of different properties.

What properties a page has is defined by its page type.

Think of a page type as a class in code. When an editor creates a new page he or she selects what page type the new page should be of. After doing so a new instance of the class is created. The class defines what properties the page has. The editor can enter values into those properties and then save the page to the database.

In object oriented programming a class acts as a template for objects. The class defines the characteristics of objects created from it. A page type isn't a class. It's an entity stored in EPiServer's database and it can be modified in admin mode. However page types are so closely resembling classes that it's possible, and indeed the preferred way, to create page types by creating classes. When doing so pages are returned as instance of those classes.

Being able to create page types in code is a new feature in version 7 of the CMS. However,

for versions 5 and 6 there was, and is, an open source project called Page Type Builder[1] that

accomplishes the same goal. As defining page types in code is the preferred approach of most EPiServer developers it's also the main approach used in this book.

Content types

It's convenient to think of pages and page types as those concepts those are fairly concrete. However, as we'll soon see, EPiServer 7 supports other type of content than pages. Therefore there's a more general concept than page types - content types.

IContent

Prior to version 7 EPiServer only supported a single type of content apart from uploaded files - pages.

Therefor, when building sites with older versions of the CMS pages were used by developers for the things than just web pages.

For instance, pages of a certain type could be used as data containers for widgets used on other pages.

In version 7 EPiServer CMS' API is no longer limited to just pages, it supports *content*. Such content can be just about anything as long as it's a class that implements the IContent interface. In other words, if you have a an instance of a class that implements IContent EPiServer's API can be used to save it to the CMS' database.

Obviously pages are content and the PageData class implements IContent.

The IContent interface, located in the EPiServer. Core namespace, has six members that a class that implements it must define:

Name : string

All content must provide a name that the CMS can use when displaying it in edit mode. The Name property can of course also be used when rendering the content for public visitors but that's up to site developers.

The PageData class implements the Name property by returning the value of the page's PageName property.

ContentLink : ContentReference

Should return a ContentReference object that uniquely identifies the content, optionally with a version, on the site. See the explanation of the ContentReference class below.

ParentLink : ContentReference

Should return a ContentReference for the content item's parent in the content tree. For page's this means the page's parent page in the page tree. However, in practice all content reside in the same tree whereas

the page tree is a view of that tree.

IsDeleted : bool

If this property returns true the content is "soft deleted", meaning it hasn't been removed from the database but has been moved to the Trash.

ContentTypeID : int

The ID of the content's content type.

ContentGuid : Guid

While the ContentLink property can be used to uniquely identify a content item on the site where it resides there may very well be some other EPiServer site somewhere that has a content item with the same ContentReference. That's not a problem, until you try to move the content to that site, for instance

using EPiServer's import/export functionality.

In order to support functionality for moving content between different EPiServer sites all content must also have a Guid, in the form of this property, that can be used to identify it.

ContentReference and PageReference

When an object implementing IContent is saved to EPiServer database it's inserted into a table with a numeric primary key. In other words, there's an unique integer ID for each content item, such as a page, stored in EPiServer's database.

This may lead us to believe that each page or other content in a site can be identified using a numeric ID. However, EPiServer supports plugging in other data sources in addition to it's own database using a concept called *content providers*. This means that while each content in EPiServer's own database is guaranteed to have a unique numeric ID that's not necessarily true for all content that can be retrieved using the CMS' API. There may very well be two, or more, contents that have the same numeric IDs if they are retrieved from different content providers. Therefore another component is necessary to uniquely identify a given content item, the provider name.

Furthermore we may want to retrieve a specific version of some content item. Or, we may want to implicitly retrieve the default version, meaning either the published version or the last version if there is no published version.

In order to address these issues EPiServer has a class named ContentReference located in the EPiServer.Core namespace. Not surprisingly the ContentReference class has three properties for identifying content – ID (int), ProviderName (string) and WorkID (int). The latter, if specified, identifies a specific version.

Heads up!

Unfortunately it's not uncommon to see developers use just the ID property of

ContentReference/PageReference objects in their code. Such code is treacherous as it may

seemingly work fine for a long time, only to later produce weird bugs when a content provider

is introduced on the site.

Always use ID + ProviderName to identify content.

PageReference

In older versions of the CMS there wasn't the concept of generic "content", there was only the more specialized concept of pages. The class for unique identifiers for pages was called PageReference. In EPiServer 7 the ContentReference class was introduced making the PageReference class seem redundant.

However, the PageReference class is still around. It inherits ContentReference and many of it's members has been moved up to the ContentReference class.

The PageReference class was partly kept for backwards compatibility reasons. However, in addition to those it also serves a purpose of being more specific than the ContentReference class. A ContentReference can be a reference to a content item of any type. A PageReference is a ContentReference but it should only be a reference to a page (a PageData object).

ContentReference members

In addition to the data bearing properties ID, ProviderName and WorkID ContentReference objects also have a few other useful members that often comes in handy:

Equals(Object) : bool

The ContentReference class overrides the Equals method inherited from Object to compare ContentReference objects by evaluating whether ID, ProviderName and WorkID all match. The ContentReference class also overloads the equality operator (==) with an implementation that invokes the Equals method.

CompareToIgnoreWorkID(ContentReference) : bool

Compares a ContentReference instance to another instance like the Equals method. However, this method only checks ID and ProviderName for equality. In many situations this is the desired behavior. In fact I'd say that I've used CompareToIgnoreWorkID a hundred times for every time I've used the Equals method

during regular site development.

It's common to see code such as

page.PageLink.ID == someContentReference.ID`

and

page.PageLink.ID == someContentReference.ID

&& page.PageLink.ProviderName == someContentReference.ProviderName`

in EPiServer projects.

The first one, comparing only ID is in most cases flawed and a bug waiting to happen as it doesn't take ProviderName into account. The second isn't flawed but clearly using more code than what is needed seeing as we have the CompareToIgnoreWorkID method at our disposal. So, a tip for writing better and shorter code is to **remember the CompareToWorkID method**.

ToString() : string

The ContentReference class overloads the ToString method inherited from Object. Its own version returns a string containing the ID property and optionally WorkID and/or ProviderName if those are specified.

This is useful for logging and debugging purposes. It's however even more useful in combination with the parsing methods described below, allowing easy serialization and de-serialization of ContentReference objects.

It can be worthwhile to know how the ToString method works in order to easily read

ContentReference objects in string form.

The return value always starts with the ID property. Then, if WorkID isn't 0, meaning that it has a value, it's appended to the return value prefixed with a single underscore. Finally, if ProviderName isn't null it's appended to the return value prefixed by two underscores.

In addition to the instance members above, the ContentReference class also provides a number of useful static members. These are listed below.

IsNullOrEmpty(ContentReference) : bool There are situations where a ContentReference can be empty,

meaning that the ID and WorkID properties are zero and the ProviderName property is null. As it's common to check whether a ContentReference variable is either null or empty, or rather validating that it's neither, the static IsNullOrEmpty method often comes in handy.

Parse(string) : ContentReference The static Parse method creates a ContentReference object from a string assuming the format that the ToString method uses. This provides a convenient way to convert back and forth between ContentReference objects and strings.

```
var contentLink = new ContentReference(42);
var complexLink = contentLink.ToString();
```

THE MISSING MANUAL

```
var parsedLink = ContentReference.Parse(complexLink);
contentLink == parsedLink; //True
```

If the argument to the Parse method is null or in any way invalid the method throws an exception of type EPiServerException.

TryParse(string, out ContentReference) : bool

Like the Parse method the TryParse method offers a way to convert a string to a ContentReference

object. However, as opposed to the Parse method the TryParse method doesn't throw an exception if the input parameter is invalid. Instead its return value indicates whether the parsing succeeded and the parsed ContentReference is returned in the form of an out parameter.

StartPage : PageReference

The start page on an EPiServer site is important in many ways. It's the page that both public visitors and editors first see when they visit the site, unless of course they are directly accessing some sub-page.

For developers it's significant as the root page when building navigation components and it's often also used to hold properties used for site-wide settings.

The ID of the site's start page is configured in a configuration file. The static StartPage property of the ContentReference class provides a convenient way for developers to get a hold of a reference to the start page.

RootPage : PageReference

Similar to the StartPage property the static RootPage property returns a reference based on configuration.

Instead of a reference to the site's start page the RootPage property returns a reference to the root page, at the very top of the page tree.

WasteBasket, SiteBlockFolder, GlobalBlockFolder

In addition to the StartPage and RootPage properties the ContentReference class has a number of other static properties that return references to, for the CMS, significant content items. One such example is the

WasteBasket property that returns a reference to the waste basket, which is a special page under which"soft deleted" content is moved to.

Renderers/templates

After defining a content type, such as a page type, editors can create content of that type and save it to EPiServer's database. However, for a public visitor to be able to see the content, and for editors to be able to edit it using On Page Editing the CMS needs to know how to present it. That is, it needs to know how to "convert" a content item into HTML.

In older versions of the CMS there was a single way to render each each page type. Page types had a property configurable in admin mode that specified a path to an ASPX file, an ASP.NET Web Forms page.

Such files were referred to as templates.

EPiServer 7 is more advanced, and complex, and allows multiple ways to render a single content item depending on context. A page may for instance be rendered as a stand-alone page, as part of another page, as a stand-alone page in a specific channel (a concept allowing tailoring of the site for specific devices or contexts) or as a part of another page in a specific channel. Therefore the terminology when building EPiServer sites has been expanded to include the word "renderer". A renderer is a component,

such as a Web Forms page or user control or an ASP.NET MVC controller or partial view that can take a content item and return HTML.

EPiServer development terminology still includes the word "template" as well and "template" and"renderer" are sometimes used interchangeably. For now, just keep in mind that both templates and

renderers are components that render HTML for a specific content item. As you'll soon see, building these components are a major, and often the largest, part of developing an EPiServer CMS site.

Data Access API

In earlier versions of the CMS its API featured a class called DataFactory. DataFactory implemented the singleton pattern, meaning that developers didn't instantiate it themselves but instead used the syntax DataFactory.Instance when needing an instance of it.

DataFactory provided methods and events for just about any operation that a site developer *and* EPiServer's own developers needed when working with PageData objects. In other words, using DataFactory one could get a page by ID (PageReference), save new and existing pages and retrieve lists of pages such

as all pages who had a specific parent. By adding handlers to various events developers could also write code that got notified whenever something happened to a page, such as it being returned or saved.

The DataFactory class provided a fairly convenient API for site developers. As a site developer needing to do anything programmatically with one or more pages one instinctively knew to use one of DataFactory's methods. Therefore it was well liked by developers. However, EPiServer also received a fair amount of

criticism due to it. Part of the criticism was that the class was bloated. However, more loud criticism was

voiced because it was very hard to isolate code from DataFactory. As each of it's methods was defined only within the class, not in an interface, it caused problems when trying to write flexible code that didn't rely on concrete implementations but instead on abstractions.

When EPiServer's developers set out to create version 7 of the CMS they decided to address this issue.

However, they also faced two other issues. First, the CMS was no longer only going to work with pages (PageData objects) but instead the more general content concept (IContent). Second, they wanted the new version to be as backward compatible as possible, making it easy to upgrade EPiServer 6 sites to the new

version.

To solve these problems they added a number of interfaces to the API, such as:

• IContentLoader - Defines methods for retrieving content items.

• IContentRepository - Inherits IContentLoader and adds methods for modifying content items.

• IPageCriteriaQueryService - Defines methods for querying for pages using criteria such as that a certain property has a specific value.

• IContentEvents - Defines events for when content are retrieved or modified.

In these interfaces they defined the methods that were needed for the functionality for the new version.

For instance, while DataFactory had a method named Get Page that returned PageData objects the IContentLoader interface defines a generic Get<T> method where the type T must be a content type.

They then made the DataFactory class implement these interfaces while keeping the old methods. Finally they added an inversion of control container through which developers could retrieve objects that implemented, amongst others, the above interfaces. The container is exposed in various ways, one of them being through a singleton class named ServiceLocator which has a Current property exposing an instance of it.

Inversion of Control (IoC)

Inversion of control is a programming technique in which object coupling is bound at runtime instead of at compile time. Imagine the following at compile time:

```
┌─────────────────┐                    ┌─────────────────────┐
│        X        │ ─────────────────▶ │          Y          │
└─────────────────┘                    └─────────────────────┘
```

Here X uses, and is bound to Y, in the code at compile time. Now, instead imagine the following:

Here X is only bound to an abstraction, I, that it knows can perform the services of Y. However, as X isn't directly coupled to Y it's possible to swap it out for something else that implements I. Of course, for this to work X needs to be able to somehow resolve, get a hold of an instance of I.

That can be achieved using Dependency Injection, where X requires an instance of I in its constructor, forcing consumers of X to supply an I. It can also be achieved using a Service Locator, in which X knows about a service, the Service Locator, which it asks for an instance of I. For both solutions it's common to use what's known as

an IoC container, an object that maps between abstractions (I in our case) and concrete implementations (Y in our case.).

By default, the implementation for all of the above interfaces registered with the container is the DataFactory.Instance singleton. This means that there are several ways to invoke the Get<T> method.

One way is by using the DataFactory singleton:

var startpage = DataFactory.Instance.Get<PageData>(ContentReference.StartPage);

Another is to retrieve an instance of the IContentLoader interface:

var contentLoader = ServiceLocator.Current.GetInstance<IContentLoader>();

var startpage = contentLoader.Get<PageData>(ContentReference.StartPage);

So, which way should we use? The first way, using DataFactory.Instance is clearly more convenient.

However, there are a couple of reasons to favor the second way.

First of all the second way, having developers rely on abstractions rather than concrete implementations, would probably have been the only way to use EPiServer's API if it hadn't been for backwards

compatibility reasons. As such, using the second approach means using the approach favored and recommended by EPiServer's developers. It's also the more future-proof approach.

Second, by using EPiServer's IoC container to resolve the abstraction that we need we make our code more flexible. Imagine for instance that we wanted to implement logging for every time someone asked the API for a page whose name started with the letter A. We could then create a custom class that implemented the relevant interfaces by wrapping the default implementation (DataFactory) and register that as the default type to use in the IoC container. Of course, that wouldn't work for when our own code accessed the API if our code bypassed the container and went straight to DataFactory.

Finally, while using the ServiceLocator class to retrieve objects through which we can get and modify content is more verbose, in practice we will often inherit from some base class provided by EPiServer when building sites and many of those offers shortcuts to the IoC container, making the code more succinct.

In this book we'll only use the second approach and won't use DataFactory.Instance. There may however be situations where EPiServer's API for working with pages is referred to a DataFactory. It's also good to know about the DataFactory class in case you run into a site built on an earlier version of the CMS.

Let's take a look at some of the most important methods of EPiServer's API for working with pages that we'll use when building sites.

IContentLoader.Get<T>(ContentReference) : T

Retrieves a content item of type T represented by the provided reference. The content is either retrieved from the database or from cache.

IContentLoader.GetChildren<T>(ContentReference) : IEnumerable<T>

Retrieves all children of type T of the content item represented by the provided reference. In other words, the GetChildren method returns all content items that are of the type T (or a sub type of T) whose ParentLink property matches the ContentReference argument.

Example input and output for the GetChildren method

IContentLoader.GetAncestors(ContentReference) : IEnumerable<IContent>

Returns all content items that forms the path to the content represented by the provided reference in the content tree.

Example input and output for the GetAncestors method

Summary

In this chapter we've looked at the most important classes and interfaces in EPiServer CMS' API, as well as some important development concepts. Perhaps most significant of all is the IContent interface as that is the definition of a type that can be stored in EPiServer's content tree.

Also important is the concept of identity where we've looked at the ContentReference and PageReference classes. We've seen that a ContentReference represents a unique identifier for something that implements

IContent. We've also seen that PageData, the class that represents a web page on the site in terms of editorial content and settings, implements IContent. While PageData objects can indeed be identified using ContentReferences the CMS' API also features a class named PageReference. PageReference

inherits ContentReference but is only used to identify pages, and certain methods in the API that work only with pages may have PageReferences as parameters instead of ContentReferences.

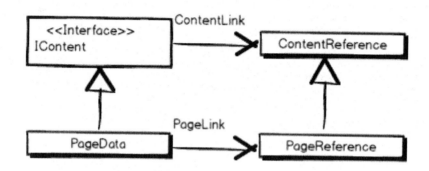

The (simplified) content and identity model in EPiServer's API.

We also discussed how to access IContent objects using the IContentLoader interface. When needing to save, move or delete content objects we need an instance of the heavier IContentRepository interface.

In order to retrieve objects that implements either of these interfaces we can either use the GetInstance method of ServiceLocator.Current or use DataFactory.Instance, although the former approach is recommended.

Building an EPiServer site

In this quick start chapter you'll learn how to create an EPiServer site from scratch – setting up a project, defining page types with properties and rendering them. After the chapter is done we'll have created a simple but functional EPiServer CMS site. The site will have two page types with templates and navigation components making it possible for visitors to browse the site based on its structure in the page tree.

The goal of this chapter is to get up and running with, and get a feel for, EPiServer site development.

Therefore we'll focus on getting things done rather than the details, leaving closer looks at the specific development concepts that we use for later chapters to cover.

Creating a site project

There are two ways of creating an empty EPiServer site suitable for further development, either by using EPiServer's Deployment Center or by using EPiServer's Visual Studio integration.

When using Deployment Center follow the same steps as for creating an Alloy sample site described in chapter two, only don't check the "Install Alloy Sample Site" checkbox.

In order to create an empty site using the Visual Studio integration select **File → New Project** in Visual Studio and select either "EPiServer Web Site" or "EPiServer Web Site (MVC)". Doing so creates a site

using the same process as Deployment Center, you even see the same progress bars. However, there are no configurations to be made, everything is pre-configured.

Both ways of creating a site accomplishes roughly the same thing, but there are some differences. When using Deployment Center you get to choose some settings such as the UI path, database name etc.

Deployment Center creates a site in the local IIS server. The Visual Studio integration doesn't. On the other hand the Visual Studio integration does a couple of things that Deployment Center doesn't. One of those things is quite minor - it sets the site's start page to the root page in episerver.config. When using Deployment Center the value for that setting is 0, which is an ID for a page that doesn't exist. As a consequence, we see an exception messages pointing that out when viewing the site and have to manually

change the setting to the root page in order to be able to proceed.

The other thing that the Visual Studio integration does that Deployment Center doesn't is far more important. It creates a Visual Studio project. Using Deployment Center you get a site with everything needed to run it, but you don't get a project in which you can place your own code, meaning that you'll have to create a project of the correct type yourself.

As most templates for web projects in Visual Studio comes with some content, such as a web.config file, you can't create the project directly in the web root of the site or you'll overwrite files that Deployment Center put there. So, when creating a site from scratch using Deployment Center we have to jump through some hoops to get a Visual Studio project set up for it.

For the reasons mentioned above, and for the convenience of just selecting **File → New Project** in Visual Studio I tend to recommend using the Visual Studio integration when setting up a new development

project. That's what we'll do now, taking our first steps to creating an EPiServer site.

Creating a site using the Visual Studio integration

Fire up Visual Studio and (given that you've installed the Visual Studio integration) select **File → New**

Project. Navigate to the EPiServer group of project templates (located under Installed/Templates/Visual

C#/EPiServer) and select **EPiServer Web Site (MVC)**. For project name use "FruitCorp.Web" and make the solution name "FruitCorp". You are of course free to choose a different project name should you want

to, but doing so may mean you'll have to alter namespaces when typing in/copying code from the book.

At the top of the dialog it's possible to choose a specific .NET version. Make sure it's either .NET Framework 4 or 4.5. The below image shows how the dialog should look. When it does, press the OK button

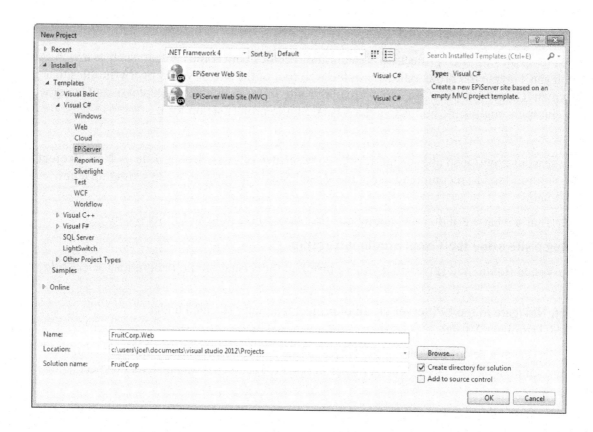

Once the project has been created you should see a project structure like in the image below in Visual Studio's Solution Explorer (press CTRL+ALT+L if you don't see Solution Explorer).

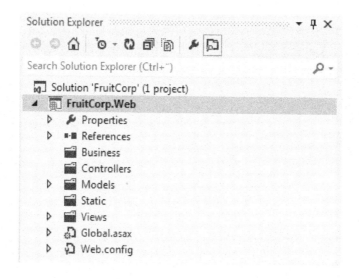

Opening up the newly created projects folder in Windows Explorer (easily done by right clicking on the project in Solution Explorer and selecting "Open Folder in File Explorer") reveals a bit more files and folders than what's shown in Visual Studio.

Server instance. The database will have the same name as the project but prefixed with "db". Now that we've got ourselves a site we should run it and take a look at it. While the Visual Studio integration doesn't create an IIS site like Deployment Center does we can use the built in development server in Visual Studio to do that instead. Simply select **Debug → Start Without Debugging** or press CTRL+F5. Doing so starts the site in the development web server and opens up its start page in the computers default browser.

After starting the site we should see EPiServer's login dialog in the browser. That may appear odd given that the browser opened up the site's start page which should be public. However, recall that a site created with the Visual Studio integration has its start page set to the site's root page (there isn't yet any "real" page to set it to) and the root page is configured, in terms of access rights, not to be shown to public visitors.

So, we're not seeing the login dialog because we asked to log in to EPiServer's UI but rather because the page that we're requesting is protected.

Log in using your Windows user name and password and you should see the root page. After doing so, let's go see how things look in edit mode. To do that we need to get to Online Center which is located under the UI path, which the Visual Studio integration has set to /episerver.

In other words, type in /episerver after the current URL in the browser and hit enter.

Once in Online Center navigate to edit mode. Opening up the Pages gadget and looking at the page tree reveals a site that is completely empty.

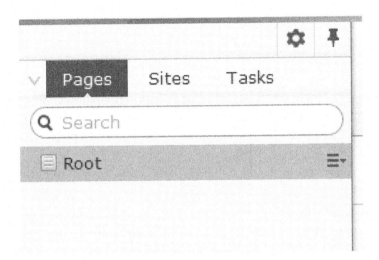

If we try to remedy this by creating a new page below the root page we find that we can't. While the dialog for creating a new page shows up and we can enter a name for the page there aren't any page types to choose from.

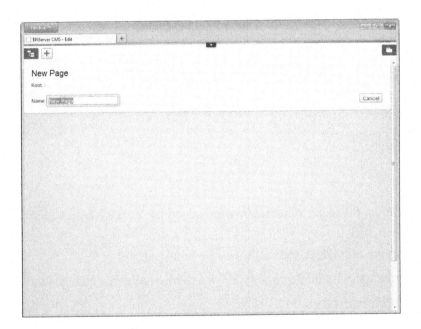

Defining a start page

With a functional but completely empty site whose start page is the root page the first natural step is create a real, albeit simple, start page. In order to do that we need to:

1. Create a page type.

2. Create a page of that type.

3. Re-configure the site to use the new page as its start page.

Let's create ourselves a page type! Create a new C# class in the Models folder by right clicking on it in Solution Explorer and selecting **Add → New item…** in the menu that pops up. Then select "Class" (located under Installed/Visual C#/Code). Name the new file "StartPage.cs" and click add. The code in the newly created file looks like this:

```
using System;

using System.Collections.Generic;

using System.Linq;

using System.Web;

namespace FruitCorp.Web.Models.Pages

{

public class StartPage

{

}

}
```

The first four lines of code, the using statements, were added by Visual Studios template for classes. At

the moment all of them are redundant and can safely be removed.

In order to turn this ordinary class into a page type we need to make two changes to it. First of all we

make it inherit from EPiServer's PageData class.

```
1 using EPiServer.Core;

2

3 namespace FruitCorp.Web.Models.Pages

4 {

5 public class StartPage : PageData

6 {
```

7 }

8 }

Compared to the first version of the class the above code has three changes:

1. The original using statements have been removed.

2. On line 1 a using statement for the EPiServer.Core namespace (where the PageData class resides) has been added.

3. On line 4 the class is made to inherit from PageData.

Now we've got a class that if instantiated would result in objects representing pages in the CMS, although as we'll see later we should never instantiate content objects ourselves. However, there's one more thing we need to do in order for EPiServer to recognize the class as a page type, meaning that it can be saved using EPiServer's API without errors and that it shows up in the list of available page types when creating a new page. We need to annotate it with the ContentTypeAttribute class, located in the EPiServer.DataAnnotations namespace.

```csharp
using EPiServer.Core;
2 using EPiServer.DataAnnotations;
3
4 namespace FruitCorp.Web.Models.Pages
5 {
6 [ContentType]
7 public class StartPage : PageData
8 {
9 }
10 }
```

That's it! We have now created a class that defines a page type, a "page type class". In order to do so we have:

1. Created an ordinary C# class.

2. Made it inherit the PageData class (line 1 + 7).

3. Added a ContentType attribute to it (line 2 + 6).

If the site is still running in IIS Express compile the project using **Build → Build Solution** or CTRL+SHIFT+B. If not run the site (which automatically compiles the project) using **Debug → Start**

Without Debugging or CTRL+F5. Now, let's try to create a page again.

This time around the list of page types to choose from in the dialog for creating a new page isn't empty, it contains a page type named "StartPage", the name of our class. Fill in "Home" as the name of the page and click on the page type to create a page.

Doing so creates a page of the StartPage type and the new page is opened up for editing. However, as opposed to when creating a new page on the Alloy sample site we're not seeing the page in on-page editing mode but in forms editing mode.

The newly created page in forms editing mode.

That's because EPiServer has no way of rendering the page as we have yet to create a template for it. We'll get to that, but for now just publish the page using the button in the top right corner, next to the blue arrow that is pointing out that it's not yet published.

Next, note the page's ID. The ID is displayed, along with the type, in forms editing mode. It's also possible to see the ID by looking at the URL in the browser when editing a content item.

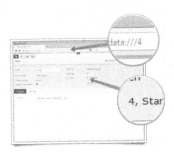

Two ways of seeing a page's ID.

With the page's ID noted head back to Visual Studio an open up episerver.config (located below web.config in Solution Explorer). In episerver.config you'll find a siteSettings element. That element has an attribute named "pageStartId" whose value currently is "1". Change that to the ID of the newly created page (most likely 4) and save the file.

Now, go back to the site and take a look at the page tree. There's no need to compile the project as the configuration file isn't compiled and the site will automatically restart upon making a change to it.

Open up the Pages gadget and note that the page now has a house icon, indicating that it is the start page for a site.

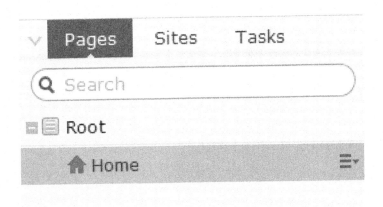

The page tree after creating a page and configuring it as start page.

THE MISSING MANUAL

Creating a template

We now have a start page but no way of rendering it. We saw that as we were taken directly to forms editing in EPiServer's edit mode after having created it or later opened it for editing.

If we were to look at the site outside of edit mode (by going to http://localhost:<some port number>/) we'd get a 404 Not Found error page.

To remedy that we'll create an initial, bare bones template for the start page. When building EPiServer site's with MVC a template for a page type corresponds to a controller and one or more views. Let's create a controller!

Right click on the Controllers folder in Solution Explorer in Visual Studio and select **Add →
Controller**.

Enter "StartController" as the controller name and leave all other settings as is ("Empty MVC Controller"as template under Scaffolding options) and click Add. The result should look something like this:

```
1 using System;

2 using System.Collections.Generic;

3 using System.Linq;

4 using System.Web;

5 using System.Web.Mvc;

6

7 namespace FruitCorp.Web.Controllers

8 {

9 public class StartController : Controller

10 {

11 //

12 // GET: /Start/

13

14 public ActionResult Index()

15 {

16 return View();

17 }

18

19 }

20 }
```

The default controller template in Visual Studio adds a few things that we can safely remove - all but the last using statements (line 1-4) and the comments (line 11-12). Now, let's turn this ordinary ASP.NET

MVC controller into a controller for EPiServer pages of our StartPage type. In order to do that we make it inherit from the generic base class PageController<T> found in the namespace EPiServer.Web.Mvc with our page type class as the type parameter. After doing so our controller looks like this:

```
using EPiServer.Web.Mvc;
2 using FruitCorp.Web.Models.Pages;
```

```
3 using System.Web.Mvc;

4

5 namespace FruitCorp.Web.Controllers

6 {

7 public class StartController : PageController<StartPage>

8 {

9

10 public ActionResult Index()

11 {

12 return View();

13 }

14

15 }

16 }
```

PageController<T> implements EPiServer's IRenderTemplate<T> interface. IRenderTemplate<T> is a marker interface (it has no members) that the CMS uses to identify classes that claim to be able to render content of a specific type. As such, we have just created a controller that EPiServer will try to use when serving requests for pages of our StartPage type.

However, the controller doesn't yet do anything useful. Our next step is to implement the Index action method so that it gets a hold of the page object for which it is being invoked, and pass it along to a view.

Developers used to the MVC pattern may find it strange that we don't create a separate view model to pass to the view. Indeed we could, but for now we're keeping it simple and we'll discuss view models later on.

Anyhow, in order to implement the Index action method we add a parameter of type StartPage named "currentPage" to it. Note that *the exact spelling and casing of the parameter name here is important.* If misspelled the parameter will be null.

```
public ActionResult Index(StartPage currentPage)

{

return View(currentPage);

}
```

Now our action method will be invoked with the StartPage object that corresponds to the page a visitor has requested. This happens through model binding in MVC, a feature of the ASP.NET MVC framework that maps HTTP request data to a model object automatically.

Under the hood: Model binding content objects

Normally the MVC framework only utilizes the data associated with the request (such as the URL, post data etc) and wouldn't be able to provide a complex PageData object with values fetched from the database. However, EPiServer handles this for us by extending the frameworks standard model binding allowing for content objects to be model bound. This way simple information from the request such as the URL is mapped to the id of a page and then conveniently translated into a complex PageData object,

fetched using EPiServer's API, by EPiServer's custom routing and model binding.

The component that identifies action method parameters that should be bound to content objects is the EPiServer.Web.Mvc.ContentDataValueProvider class which looks for parameters named "current Data", "currentPage" and "currentBlock". In other words, in order for the above action method to work

it's vital that the parameter has exactly one of those three names, and here "currentPage" is clearly the most descriptive.

Now our simple action method handles StartPage objects and passes them along, as model objects, to a view. However, we have yet to add such a view. In order to do right click somewhere within the code for the method and select **Add View**. Let the view name be "Index" and check the "Create a strongly-typed view" checkbox. Type in StartPage as the Model class. If the type doesn't appear in auto complete or in the dropdown in the dialog it's because you haven't compiled the project yet. If so, just type in the name

of the class and adjust the namespace in the @model directive after the view has been created. Finally, un-check the "Use a layout or master page" checkbox (we'll modify it to use a layout later) and click the Add button.

The code for the resulting view looks like this:

```
1 @model FruitCorp.Web.Models.Pages.StartPage
2
3 @{
4 Layout = null;
5 }
```

6
7 *<!DOCTYPE html>*

```
9 <html>
10 <head>
11 <meta name="viewport" content="width=device-width" />
12 <title>Index</title>
13 </head>
14 <body>
<div>
16
17 </div>
18 </body>
19 </html>
```

Add some hard coded greeting message inside the div tag (on line 16), like the code below and then either start the site if it's not still running, or compile the project if it is. Note that if you start the site while having a view open the site will be opened with the URL of the view. This can sometimes be confusing as the view isn't meant to be viewed stand-alone, without first going through a controller and you may see some strange error message. If so, just ignore that and type in the URL for edit mode in the browsers navigation tool bar.

```
<div>
<h1>Hello world, and welcome to our site!</h1>
</div>
```

With this very simple view in place compile the project and take a look at the site. You'll find that when viewing the start page in edit mode you're no longer automatically thrown directly into forms editing mode. Also, viewing the site outside of edit mode no longer returns a 404 page but instead a page corresponding to the view you just created.

It may not be the prettiest thing ever seen in a browser, but we now have a functional view. So far we haven't rendered any properties though. What little is shown on the page is entirely hard coded. We'll soon look at how to define properties in page types, but before we do that we can render one of the "builtin" properties inherited from the PageData class, the

PageName property. To render the page's name in the view remove the hard coded greeting in the H1 tag and replace it with @Model.PageName, like this:

<h1>@Model.PageName</h1>

Save the view and reload the start page either in EPiServer's edit mode or publicly on the site. (There's no need to compile the project as views aren't compiled, or rather they are compiled Just-In-Time.)

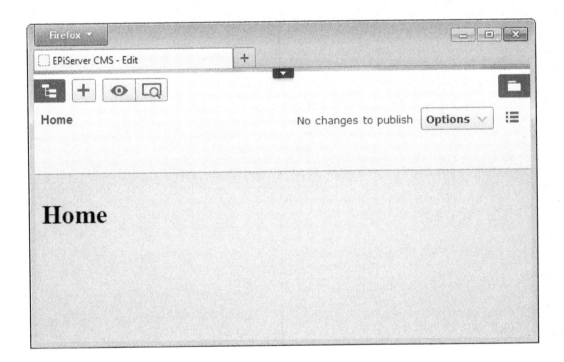

Now the view is displaying a property from its model object, here meaning the PageData object for the currently viewed page as we're passing it directly to the view. Should we change the page's name in forms edit mode the change will be reflected in the view. However, we're not able to edit the PageName property

in on-page-edit mode. There are several ways to accomplish that, which we'll look at later on, but for now we can make it happen using perhaps the simplest approach; with EPiServer's PropertyFor HTML helper method.

In its simplest form the PropertyFor method (located in the EPiServer.Web.Mvc.Html namespace) takes a lambda expression as its single parameter. The expression is of type Func<TModel, TValue> meaning that the expression should be a function that receives an object of the same type as the views model and returns a value of any type. Using this we tell the method the name of the property we

THE MISSING MANUAL

want to render in a strongly typed way. To use it to render the PageName property, update the code inside the H1 tag to look like this:

```
<div>
```

`<h1>@Html.PropertyFor(x => x.PageName)</h1>`

`</div>`

After saving the view (and possibly reloading the start page in edit mode) we can now edit the PageName property, which we currently use as the page's heading, in on-page-edit mode.

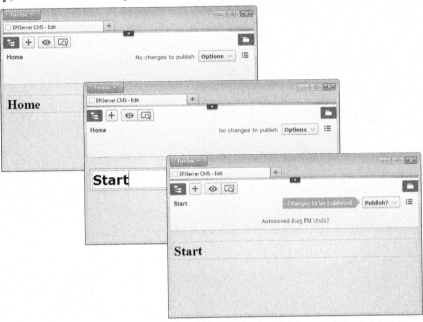

The page name editable, and being changed, in on-page-edit mode.

Congratulations! We have just created, or seen how to create, a page type and a template. Albeit extremely minimalistic, we have just gone through how to create a fully functional EPiServer CMS site. To make the site a bit more useful we'll now create a second page type and add some properties of our own to

it. However, prior to doing that we should make it easier for ourselves to work with views by adding a layout. We'll also add some CSS that will make it easier to position elements in views and make them look better.

.

Web Forms note

When using Web Forms there is no such concept as controllers and views. Instead we create pages. A page is made up of two components, a file with an .aspx file name extension containing HTML markup and various types of controls and a code-behind file with an .aspx.cs file name extension (when using C#) containing methods and properties. When an ASPX file is requested on the web server for the first time it's JIT compiled with the class in the code-behind file (which is already compiled) as its base class.

When building an EPiServer site with Web Forms the equivalent of creating a template as a controller inheriting from PageController<T> and one or more views in MVC is to create an ASPX page and make it inherit from a class named TemplatePage found in the EPiServer namespace. As with the PageController class, the TemplatePage class implements IRenderTemplate<T> and has a type parameter (although there is one without it as well for legacy reasons) with which we tell EPiServer what types of pages the template handles. When creating such a template we can retrieve the specific page that it's supposed to render using the CurrentPage property inherited from the TemplatePage<T>

class. The CurrentPage property will be of type T.

While the two don't work exactly the same the rough equivalent of the PropertyFor method in MVC is a server control simply named Property found in the namespace EPiServer.Web.Controls.

The Property control has a string property named PropertyName with which we tell the control the name of the property that we want it to render from the current page (the CurrentPage property in the host ASPX page). The Property control also has a number of other properties that we may optionally use to customize its rendering. For instance, the Property control always wraps the value of the property in a HTML element of some sort depending on the property's type and its CustomTagName property can be used to override the default behavior and render the property's value within a specific tag.

In order to render the PageName property the same way as we did in the MVC view we would in Web Forms have written:

```
<div>
<EPiServer:Property PropertyName="PageName" CustomTagName="h1" runat="server" />
</div>
```

Layout and CSS

In the previous section we created a simple template for a page type in the form of a controller and a view.

As soon as we add a second page type we run in to a maintainability problem with regards to views. Many of the elements are often the same on all or most of a sites views. For instance, all views for pages when they are rendered as full-scale pages (as opposed to as partials on other pages) need to have HTML and

BODY tags. Also the content, or functionality to generate content, in both the HEAD and BODY part of them may be the same. Apart from addressing the code duplication problem, having some parts of views located in a common place help to create and maintain a consistent look and feel for them.

When using ASP.NET MVC and the Razor view engine, as we are, this problem is easily solved using something called layouts. A layout looks almost like a view, but it's not "complete" and can't be used on its own. Sort of like an abstract class it works as a template for other views. A layout must somewhere in its markup invoke a method called RenderBody. Views that use a layout define the content that is inserted in place of the RenderBody call. It's also possible for layouts to define other sections that views can insert content into.

To make our existing view use a layout we'll start by creating the layout. Right click on the folder named"Shared" inside the "Views" folder in Solution Explorer and select **Add → View…**. Name the view "_Root" (by convention layout names a prefixed with an underscore). Un-check the "Use a layout or master page"

checkbox and click the Add button. The resulting layout file should look something like this:

```
@{
Layout = null;
}
<!DOCTYPE html>
<html>
<head>
<meta name="viewport" content="width=device-width" />
<title>_Root</title>
</head>
<body>
<div>
```

```
</div>
</body>
```

```
</html>
```

As you can see, the layout looks almost identical to how the view we created in the previous section initially looked. As such there is little work needed for us here given that our current task is to make our existing view use a layout but generate the same output. Therefore all we have to do is insert a call to RenderBody between the two DIV tags.

```
<div>
```

```
@RenderBody()
```

```
</div>
```

Next we go back to the start page's view and remove all markup except for the H1 tag in which we render

the PageName property. After doing so the view should look like this:

```
1 @model FruitCorp.Web.Models.Pages.StartPage

2

3 @{

4 Layout = null;

5 }

6

7 <h1>@Html.PropertyFor(x => x.PageName)</h1>
```

Finally we need to instruct the view to use our newly created layout. We do that by setting Layout property

to the path to the layout file on line 4, like this:

```
1 @model FruitCorp.Web.Models.Pages.StartPage

2

3 @{

4 Layout = "~/Views/Shared/_Root.cshtml";

5 }

6

7 <h1>@Html.PropertyFor(x => x.PageName)</h1>
```

That's it. Our view is now using a layout. So far we've yet to reap any benefits from that, but as soon as we create another template we certainly will. Next, let's add a style sheet that will help us with the graphical layout of the site. In order to focus on EPiServer development without getting tangled up in styling with CSS we'll use an existing CSS framework. You are of course free to use any CSS framework or custom CSS

that you'd like, both when developing sites and when following this book. However, in the examples in the book we'll be using the Twitter Bootstrap framework. I choose Bootstrap for a couple of reasons. First of all it's a popular, widely used framework. It's also a framework that seems to suit back-end oriented developers well. Last but not least, EPiServer uses Twitter Bootstrap in several of their template and demo sites.

Bootstrap 101

Bootstrap is a front-end framework built at Twitter. It consists of three components; CSS files, a font used for icons and JavaScript files. There are two CSS files, plus two that are minified versions of the first two. These are named bootstrap.css and bootstrap-theme.css. The first is Bootstrap's core style sheet while the second can optionally be used to make various elements look more three-dimensional.

Using Bootstrap's style sheet it's possible to create an entire site without ever writing a line of CSS as it provides ready-to-use CSS classes for grids, forms, tables and a host of other components. By using the JavaScript file as well it's easy to add transitions, modal dialogs, tool tips, slideshows etc without having to write any JavaScript at all.

Of course, while it's possible to *only* use Bootstrap when building a site most sites that uses it do extend it with custom styling and additional functionality in JavaScript. Nevertheless, Bootstrap provides a solid foundation to build upon and, which we use it for in this book, a simple way of creating well functioning user interfaces without *having* to write CSS.

At the heart of Bootstrap is its grid system. The grid system utilizes twelve columns. This means that we can use CSS classes from Bootstrap to create rows that can contain a variable number of "cells" where each cell can be anything from 1/12 of the total width of available (the width of the page or of the parent element) to the full width of the grid. Rows are elements, typically div tags, with a CSS class named

"row". Cells, or columns, within rows are elements, again typically div tags, with a "col-md-*" CSS class, where the "*" is a number between one an twelve. The number defines how many column widths wide it is. That is, the width of a col-md-* element is at least * times the total width pixels divided by 12.

Here's an example:

Building an EPiServer site 113

.

```
1 <!DOCTYPE html>
2 <html>
3 <head>
4 <link href="css/bootstrap.min.css" rel="stylesheet">
5 <style>
6 .col-md-8 {
7 background: red;
8 }
```

```
 9 .col-md-4 {
10 background: green;
11 }
12 </style>
13 </head>
14 <body>
```

```
15 <div class="row">
16 <div class="col-md-8">
17 <h2>col-md-8</h2>
18 </div>
19 <div class="col-md-4">
20 <h2>col-md-4</h2>
21 </div>
22 </div>
23 </body>
24 </html>
```

The above code defines a simple HTML page with Bootstrap's CSS added (assuming it exists in a folder named "css" relative to from where the page is loaded). It also contains a small piece of in-line CSS styling in order to make col-md-8 elements red and col-md-4 elements green. In the body it defines a single row with two columns, the first eight column widths wide and the other four column widths wide. Here's how it looks in a browser:

To center the grid (in the above case the single row) and make it narrower than the full page width it can be wrapped in an element with a class named "container". Also, besides adding col-md-* classes to columns it's possible to position them at a certain number of column widths left of the previous column using a class named "col-md-offset-*" where the * is the number of column widths to offset it by.

Besides the grid system Bootstrap's stylesheet also contains CSS classes for a host of components, typography etc. As an example there are classes for a number of different types of navigations.

Below is an example with a more complex grid wrapped in a container as well as one of the navigation types.

```
1 <!DOCTYPE html>
2 <html>
3 <head>
4 <link href="css/bootstrap.min.css" rel="stylesheet">
5 </head>
6 <body>
7 <div class="container">
8 <nav class="navbar navbar-inverse">
9 <ul class="nav navbar-nav">
```

```
10 <li class="active"><a href="#">Home</a></li>
11 <li><a href="#">About</a></li>
12 <li><a href="#">Contact</a></li>
13 </ul>
14 </nav>
15 <div class="row">
16 <div class="col-md-8">
17 <h2>col-md-8</h2>
18 </div>
19 <div class="col-md-4">
20 <h2>col-md-4</h2>
```

```
21 </div>
22 </div>
23 <div class="row">
24 <div class="col-md-3">
25 <h2>col-md-3</h2>
26 </div>
27 <div class="col-md-7 col-md-offset-2">
28 <h2>col-md-7 + col-md-offset-2</h2>
29 </div>
30 </div>
31 </div>
32 </body>
33 </html>
```

Rendered in a browser the above markup produces this:

For more information about Bootstrap, documentation, examples and downloads see Bootstrap's site, http://getbootstrap.com/. Also, note that what's discussed here is version three of Bootstrap. Most of the discussion also applies to version two, but in that version we used "span*" classes instead of "col-md-*"

classes.

Bootstrap can easily be added to a site using NuGet. Simply search for "twitter bootstrap" in NuGet's UI and install it or type in "Install-Package Twitter.Bootstrap" into the package manager console.

With that done, we can add Bootstrap's style sheet to our layout file. The NuGet package places Bootstrap's CSS files into the project in a folder called Content, meaning that we can link to the style sheet with <link Building an EPiServer site 115 href="_/Content/bootstrap/bootstrap.min.css" rel="stylesheet">. While we're working with our layout let's also center what ever is rendered in place of the RenderBody method by adding a "container" class to the DIV tag that wraps it. Here's what the layout should look like:

href="_/Content/bootstrap/bootstrap.min.css" rel="stylesheet">. While we're working with our layout let's also center what ever is rendered in place of the RenderBody method by adding a "container" class to the DIV tag that wraps it. Here's what the layout should look like:

```
1 @{
2 Layout = null;
3 }
```

```
4
5 <!DOCTYPE html>
6
7 <html>
8 <head>
```

```
9 <meta name="viewport" content="width=device-width" />
10 <link href="~/Content/bootstrap/bootstrap.min.css" rel="stylesheet">
11 <title>_Root</title>
12 </head>
13 <body>
14 <div class="container">
15 @RenderBody()
16 </div>
17 </body>
18 </html>
```

Web Forms note

When building sites with Web Forms, or when using the Web Forms view engine when using ASP.NET MVC, we can use "Master Pages" instead of layouts in the Razor view engine. When doing so EPiServer doesn't provide or require any particular base class for the master page.

Creating a page type with properties

We've now got a simple start page and a basic layout to work with. The start page doesn't look like much and that's natural as it's often based on other content on the site and therefore typically not the first page type and template to be completed first. Still, we needed to have some sort of start page in order to have a functioning site.

So, while we'll come back to it later, like in many real-world EPiServer projects the next natural step is to create a page type that will be more widely used across the site.

While EPiServer sites often have quite a few page types it's common to have some sort of "standard" page type used for various purposes where editors need to publish information. On a corporate web site such a page type may be used for the majority of the sites page while on a newspaper's site it may be used for "utility" pages such as "About the newspaper" while using more specialized page types for actual articles.

Anyhow, let's create a typical standard page type suitable for our site. "Standard pages" often consist of a headline, a preamble and a body of copy. To begin with, let's create the page type class.

Just as the previous time we created a page type we start by right clicking on the Models folder in Solution Explorer Building an EPiServer site 116 and select to add a new C# class. We'll name the class "StandardPage". Once created we can clean up unused using statements and perform the steps necessary for turning the ordinary class into a page type class recognized by EPiServer.

As you may recall from when we created the StartPage class there are two steps to converting a regular C# class into a page type class. We make it **inherit from PageData** and we **add a ContentType attribute**

to it. After doing so the StandardPage class looks like this:

```
1 using EPiServer.Core;
2 using EPiServer.DataAnnotations;
3
4 namespace FruitCorp.Web.Models.Pages
5 {
6 [ContentType]
7 public class StandardPage : PageData
8 {
9 }
10 }
```

Now, let's add some properties to the page type class. In it's simplest form a C# property recognized as a page type property by the CMS is an ordinary C# property with automatic/compiler generated getters and setters. There's just one little gotcha; it needs to be marked as virtual, meaning that it can be overridden in sub classes. If it isn't, the project compiles but we get an exception saying that the property needs to be virtual when the site starts up. We'll get back to the reason for this.

To start with we can use the PageName property inherited from PageData as headline but we need a preamble property and a property for the main body of copy. So, given that we want the preamble as a string (seems natural, right) we'll add a string property to the class. We'll call it MainIntro.

```
1 using EPiServer.Core;
2 using EPiServer.DataAnnotations;
3
4 namespace FruitCorp.Web.Models.Pages
5 {
6 [ContentType]
7 public class StandardPage : PageData
8 {
```

```
 9
10 public virtual string MainIntro { get; set; }
11
12 }
13 }
```

That's it. When creating a page of this type editor we'll see a property named MainIntro edited using a textbox in forms editing mode, as well as in on-page-edit mode if we render it in a template using the PropertyFor method. Next we need to add the property for the page's main content. Here we want editors to be able to insert richer content such as headings and images and to use formatting such as font modifications (bold, italic etc.). In other words we need a property that stores HTML and is edited using a WYSIWYG editor. In order to do that we create a property of an EPiServer specific class named XhtmlString, found in the EPiServer.Core namespace.

```
1  using EPiServer.Core;
2  using EPiServer.DataAnnotations;
3
4  namespace FruitCorp.Web.Models.Pages
5  {
6  [ContentType]
7  public class StandardPage : PageData
8  {
9
10  public virtual string MainIntro { get; set; }
11
12  }
13  }
```

That's it. When creating a page of this type editor we'll see a property named MainIntro edited using a textbox in forms editing mode, as well as in on-page-edit mode if we render it in a template using the PropertyFor method. Next we need to add the property for the page's main content.

Here we want editors to be able to insert richer content such as headings and images and to use formatting such as font modifications (bold, italic etc.). In other words we need a property that stores HTML and is edited using a WYSIWYG editor. In order to do that we create a property of an EPiServer specific class named XhtmlString, found in the EPiServer.Core namespace.

```
using EPiServer.Core;
using EPiServer.DataAnnotations;
namespace FruitCorp.Web.Models.Pages
```

```
{
    [ContentType]
    public class StandardPage : PageData
    {
```

```
public virtual string MainIntro { get; set; }

public virtual XhtmlString MainBody { get; set; }

}

}
```

Looks good! We now have a page type that supports a headline (by using the PageName property), a simple string based preamble and a richer main text.

Property naming conventions

While EPiServer doesn't enforce any naming policy for properties added to page types or other content types the names of the properties created above, MainIntro and MainBody wasn't arbitrary. When developing sites using Web Forms EPiServer provides a number of server controls (templated controls used to easily render lists etc). Some of those controls look for properties with specific names, most prominently MainIntro and MainBody, and assume that they are used as the main content of a page and as a short introduction respectively.

While there aren't any such controls for MVC and many EPiServer sites built with Web Forms don't use those controls many EPiServer developers still name their properties MainIntro and MainBody. Especially with regards to MainBody the name has very much become a convention in the EPiServer development community and **it is considered a best practice to name the**

main rich content property in page types MainBody.

This helps other developers to quickly identify the property.

With the page type in place it's time to create a template for it. We begin by creating a controller that handles pages of this type. In order to do that we repeat the steps performed when creating a controller for the StartPage page type. Meaning that we add a new, empty MVC controller in the Controllers folder in the project, only this time we name it StandardPageController. Next we make it inherit from PageController<StandardPage> and implement its Index action method to take a StandardPage object as parameter that it immediately passes on to the view. As before the name of the parameter is important, it

needs to be *exactly* "currentPage" (or "currentData" or "currentBlock").

Here's what the controller should look like:

```
using EPiServer.Web.Mvc;

using FruitCorp.Web.Models.Pages;

using System.Web.Mvc;

namespace FruitCorp.Web.Controllers
```

```csharp
{
    public class StandardPageController : PageController<StandardPage>
    {
        public ActionResult Index(StandardPage currentPage)
        {
```

```
return View(currentPage);
}
}
}
```

Next we add a view for the Index method by right clicking somewhere in it and selecting **Add View**.

Before clicking the Add button in the "Add View" dialog we set the model class to StandardPage. This time we also check the "Use a layout or master page" checkbox and enter (or browse to) the path to our layout file, _/Views/Shared/_Root.cshtml. The resulting view looks like this:

```
1 @model FruitCorp.Web.Models.Pages.StandardPage
2
3 @{
4 ViewBag.Title = "Index";
5 Layout = "~/Views/Shared/_Root.cshtml";
6 }
7
8 <h2>Index</h2>
```

To start with we can remove the assignment to ViewBag.Title (line 4) and the H2 tag (line 8) added by Visual Studio's view template. Next we add a simple grid with three columns to the view using CSS classes

from Bootstrap.

```
1 @model FruitCorp.Web.Models.Pages.StandardPage
2
3 @{
4 Layout = "~/Views/Shared/_Root.cshtml";
5 }
6
7 <div class="row">
8 <div class="col-md-3"></div>
9 <div class="col-md-6">
10
11 </div>
12 <div class="col-md-3"></div>
13 </div>
```

For now the columns only serve to make the main content in the page, which we'll put in the middle column, narrower than the full grid width. Later we can use the other columns for navigation and other components. Now, let's render the three properties that we want to display on the page. As before we do this using the PropertyFor HTML helper method.

```
<div class="col-md-6">

<h1>@Html.PropertyFor(x => x.PageName)</h1>

<p class="lead">
```

```
@Html.PropertyFor(x => x.MainIntro)
```

```
</p>
```

```
@Html.PropertyFor(x => x.MainBody)
```

```
</div>
```

As shown above we wrap the PageName and MainIntro properties in HTML tags but not the MainBody property. We do this as we want the values in them to be rendered inside block level elements. Also, we may want them to have a certain element type, H1 for PageName, or be able to add CSS classes, as in the "lead" class from Bootstrap for MainIntro. However, we don't wrap the MainBody property in an element.

While we of course could do that should we have a reason to XhtmlString properties are typically edited using a WYSIWYG editor (TinyMCE) which normally ensures that the content is in one or more suitable elements. For instance, TinyMCE creates P tags for each paragraph entered.

Now, let's take our new page type for a spin. Start the site, or only compile the project if it's already running, and create a page named "About Fruit Corp" under the sites start page. Make sure you select the "StandardPage" page type and not the "StartPage" one. Once created, enter some text in the properties. A filler text generator, such as Veggie ipsum[1], may come in handy.

A new standard page just after having been created.

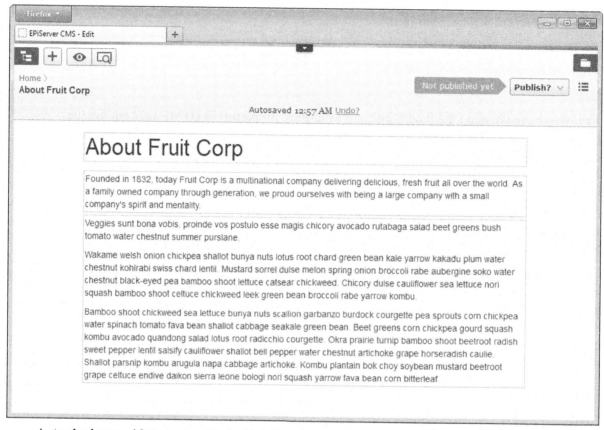

A standard page with text entered in the MainIntro and MainBody properties.

The page type and template looks pretty good. However, there are a couple of issues with the MainIntro property. First of all it's looks different in on-page-edit mode than it does when viewing the page outside of edit mode. Second, it is edited using a textbox which is a bit small. A textarea would be better. We also have other work to do before we can claim that this page type has been created with good EPiServer developer craftsmanship, such as making the property names more descriptive and translated to different languages in edit mode. We'll look at how to fix these issues and improve the quality of the page type

in upcoming chapters, when we look at properties on content types in more detail and learn more about property rendering.

Why separate properties?

One may wonder why we used three different properties for the heading, preamble and main body of copy for the page type that we just created. After all, couldn't we just have added a single HTML property and let editors create the headline and preamble with specific CSS classes inside it instead? Indeed we could have. However, using separate properties here have several benefits.

By using separate properties we help guide editors with regards to what content they can, or are expected, to enter in order to create pages that are in harmony with the site's structure and design. Also, having the heading and preamble as separate properties, and as simple strings, is handy when we may want to render them separately. For instance, the heading property should probably be repeated in the title tag (as in <head><title>) and the preamble may be used as the pages meta description. The heading and preamble is often also used when rendering the page as part of other pages, for instance in listings.

Finally, as we'll see later on, properties can be validated in order to ensure that editors enter values in them, or that entered match a certain condition. While we may not want to force editors to enter a preamble, having it as a separate property leaves that option open.

Creating a top menu

With our two page types and their templates we have what's needed to create a simple site in terms of entering content. There is however something missing for the site to be usable.

Visitors need to be able to navigate on the site. To address that we need to create a couple of navigation components.

However, to be able to see these components in action in a meaningful way we need some more content on our site.

So, let's create ourselves some pages on multiple levels. Create another page below the start page named "Products" and create a number of pages below the about-page that we created in the previous section.

Make them all using the StandardPage page type. The image below shows how the page tree should look.

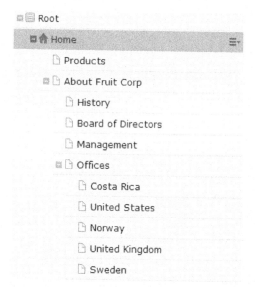

The page tree after creating some content that could be considered realistic on a corporate website.

With some content on our site, let's create ourselves a first navigation component. Most sites tend to have a top navigation. On an EPiServer site, where the page tree is often used to structure the site, it's convenient to build it so that it's made up of the pages located directly below the start page. We'll create a top menu that works in that way in the form of a HTML helper, an extension method for instances of

ASP.NET MVC's HtmlHelper class.

Begin by creating a new folder in the projects root level named "Helpers". Inside the folder add a new C# class named "NavigationHelpers".

Make the class static, as that's a requirement for being able to create extension methods in it

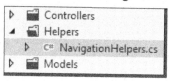

```
namespace FruitCorp.Web.Helpers
{
public static class NavigationHelpers
{
}
}
```

Next, add a static method named "RenderMainNavigation" with no return value (void). The method should have a parameter of type HtmlHelper (in the System.Web.Mvc namespace), prefixed with the this keyword, so that the method can be used as an extension method for HtmlHelper objects.

Building an EPiServer site 122

```
using System.Web.Mvc;
namespace FruitCorp.Web.Helpers
{
public static class NavigationHelpers
{
public static void RenderMainNavigation(this HtmlHelper html)
{
}
}
}
```

Before we implement the method, let's use it in our layout so that we can look at the results once we do start implementing it. Open up the layout file (Views/Shared/_Root.cshtml) and invoke the method just before the call to the RenderBody method (line 15 below). Note that as our method doesn't return anything, but instead will write directly to the response when invoked, we need to wrap the call to it in curly braces and terminate the statement with a semicolon. Also, as the method is located in one of our own custom namespaces we need to add a using statement (line one below).

```
@using FruitCorp.Web
2 <!DOCTYPE html>
3
4 <html>
5 <head>
6 <meta name="viewport" content="width=device-width" />
```

```
7 <title>_Root</title>
8 <link rel="stylesheet" href="~/Content/bootstrap/bootstrap.min.css" />
9 </head>
10 <body>
11 <div class="container">
12 <div class="row">
13 <div class="col-md-2"><h3 class="muted">AcmeFruit</h3></div>
14 </div>
15 @{ Html.RenderMainNavigation(); }
16 @RenderBody()
17 </div>
18 </body>
19 </html>
```

Instead of placing using statements for required namespaces in views, like on the first line in the code above, you can add namespaces that should be automatically imported for all views to the configuration/system.web.webPages.razor/pages/namespaces node in the web.config file located in the Views folder.

Below is an example (with the parts of the web.config file not relevant for this discussion omitted) :

```
<?xml version="1.0"?>
<configuration>
<system.web.webPages.razor>
<pages pageBaseType="System.Web.Mvc.WebViewPage">
<namespaces>
<add namespace="FruitCorp.Web.Helpers"/>
</namespaces>
</pages>
</system.web.webPages.razor>
</configuration>
```

Here in the book I'm not using this approach in order to make it clearer what methods and classes are used in views. However, in a real project I'd recommend adding the namespaces this way, otherwise the views tend to be bloated with a lot of using statements.

Now, let's create a first simple implementation of the RenderMainNavigation method. The navigation will be an unordered list (UL+LI tags) with Bootstrap's CSS classes "nav" and "navbar-nav" wrapped in a NAV element with "navbar" and "navbar-inverse" CSS classes. So, as a first step we write those elements to the response.

```
public static void RenderMainNavigation(this HtmlHelper html)
{
var writer = html.ViewContext.Writer;
```

```
//Top level elements
writer.WriteLine("<nav class=\"navbar navbar-inverse\">");
writer.WriteLine("<ul class=\"nav navbar-nav\">");
//Close top level elements
writer.WriteLine("</ul>");
writer.WriteLine("</nav>");
}
```

Return MvcHtmlString or write to the response?

Most HTML helper methods in ASP.NET MVC return a MvcHtmlString object whose string value is to the view. Some methods come in pairs where one returns a MvcHtmlString and one writes directly to the response. The latter have the same names as the former but prefixed by "Render", such as Action and RenderAction.

Methods that return a MvcHtmlString are slightly more convenient to use in views as the syntax is shorter, while those that write directly to the response may in theory be faster as they don't have to build up a string. We can use either approach when creating custom HTML helper extension. Here in the book we use the "Render" approach for long methods as that produces more concise code suitable for reading.

Depending on the requirements from site to site it may be that the start page should be included in the top menu. At least for now, when we aren't offering any other way to get to the start page from other pages, we want this behavior for our site. So, before we list any other pages we begin by writing a link to the start page, inside a LI tag, to the response.

In order to create a link to a page, complete with an A HTML tag with the href attribute set to the page's URL and the PageName property as text, we can use one of EPiServer's HTML helper methods, the PageLink method found in EPiServer.Web.Mvc.Html.StructureHtmlHelperExtensions. In its simplest

form the PageLink method takes a single parameter, a PageReference representing the page to create a link to.

To get a hold of a reference to the start page we can use the static StartPage property of the ContentReference class.

The PageLink method returns a MvcHtmlString, an object that wraps a string that has been HTMLencoded and therefore is safe to write to the response. In order to write the underlying string from that to the response we can use its ToHtmlString method.

Using the PageLink method to create link to the start page wrapped in a LI tag, the part of the method between writing the opening and closing tags for the UL element should look like this:

```
//Assuming using statements for EPiServer.Core and
// EPiServer.Web.Mvc.Html has been added to the class.
```

...

```
writer.WriteLine("<li>");
writer.WriteLine(html.PageLink(ContentReference.StartPage).ToHtmlString());
writer.WriteLine("</li>");
```

...

Next we need to write links to the pages located directly below the start page. As we learned in chapter four, we can get a hold of the children of a given page, or other type of content, using the GetChildren method defined in the IContentLoader interface.

So, as a first step we use ServiceLocator.Current to fetch an implementation of IContentLoader.

Then we call the content loaders GetChildren method with ContentReference.StartPage as argument.

We use PageData as type parameter as we're interested in all

pages but no other type of content.

```
//Assuming a using statement for EPiServer has been added to the class.
...
var contentLoader = ServiceLocator.Current
.GetInstance<IContentLoader>();
var topLevelPages = contentLoader
.GetChildren<PageData>(ContentReference.StartPage);
...
```

Next we iterate over the start page's children, writing a LI element with a link for each page, using the same approach as we used when we added wrote the link to the start page.

```
...
foreach (var topLevelPage in topLevelPages)
{
writer.WriteLine("<li>");
writer.WriteLine(html.PageLink(topLevelPage).ToHtmlString());
writer.WriteLine("</li>");
}
...
```

That's it! While we have yet to filter out pages that shouldn't be visible in the navigation or highlight the link to the page in the currently viewed branch of the page tree, we now have an otherwise working method for our top menu.

The complete class looks like this:

```
using EPiServer;
using EPiServer.Core;
using EPiServer.ServiceLocation;
```

```csharp
using EPiServer.Web.Mvc.Html;
using System.Web.Mvc;
namespace FruitCorp.Web.Helpers
{
public static class NavigationHelpers
{
public static void RenderMainNavigation(this HtmlHelper html)
```

```
{
    var writer = html.ViewContext.Writer;
    //Top level elements
    writer.WriteLine("<nav class=\"navbar navbar-inverse\">");
    writer.WriteLine("<ul class=\"nav navbar-nav\">");
    //Link to the start page
    writer.WriteLine("<li>");
    writer.WriteLine(html.PageLink(ContentReference.StartPage).ToHtmlString());
    writer.WriteLine("</li>");
    //Link to the start pages children
    var contentLoader = ServiceLocator.Current
    .GetInstance<IContentLoader>();
    var topLevelPages = contentLoader
    .GetChildren<PageData>(ContentReference.StartPage);
    foreach (var topLevelPage in topLevelPages)
    {
        writer.WriteLine("<li>");
        writer.WriteLine(html.PageLink(topLevelPage).ToHtmlString());
        writer.WriteLine("</li>");
    }
    //Close top level elements
```

Building an EPiServer site 126

```
    writer.WriteLine("</ul>");
    writer.WriteLine("</nav>");
}
}
```

Compile the project and take a look at the site, which should now contain a top menu allowing us to navigate between the site's start page its children.

THE MISSING MANUAL

HTML as strings in C#? Yuck!

The RenderMainNavigation method that we have just created writes HTML directly to the

response in the form of C# strings. This makes it quite long and not very flexible with regards to what HTML it outputs.

This is intentional here in the book as I want to keep the code as

straight forward and simple, but not necessarily as terse, as possible.

In the ASP.NET MVC version of the Alloy sample site[2], which is available for download on

EPiServer World, you'll find an HTML helper extension method named MenuList (located in the EPiServer.Templates.Alloy.Helpers.HtmlHelpers class).

This method can be used to create navigation components, ranging from top menus to tree structures to breadcrumbs with custom

templates for links. However, the flexibility comes at a price; it can only be used with the Razor view engine and its signature and implementation is quite complex. For a real project you may want to take a look at and copy the code for this method. However, for the sake of learning in this book we're keeping it simple, albeit verbose.

Highlighting the current branch

Our top menu works as it should, linking to the start page and its children, but it doesn't provide any visual indication of where on the site a visitor is. Therefore our next step is to add a CSS class, "active" from Boostrap, to a link if it links to the currently viewed page. In order to do that we need to figure out what page the current request to the site is for. In many situations EPiServer provides an easy way to do that, for instance through model binding in action methods.

However, in an HTML helper extension, which is just a regular C# method, there's no such help.

Thankfully our method takes a HtmlHelper as parameter through which we can access the current request context (System.Web.Routing.RequestContext) and EPiServer's API provides an extension method for that, named GetContentLink in EPiServer.Web.Routing.RequestContextExtension, for retrieving a

reference to the content item that the request routes to.

Using these components we can retrieve a reference to the current content in our method, like this:

//Assuming using statement for using EPiServer.Web.Routing.

var contentLink = html.ViewContext.RequestContext

.GetContentLink();

Now we can replace both places where we write an opening LI tag for a link (currently looking like this:

writer.WriteLine("<li>");) with an if-else statements that adds the "active" CSS class if the link is for the currently viewed page. Here's how the code for rendering the link to the start page looks after this modification:

if (ContentReference.StartPage.CompareToIgnoreWorkID(contentLink))

{

writer.WriteLine("<li class=\"active\">");

}

else

{

writer.WriteLine("<li>");

}

writer.WriteLine(html.PageLink(ContentReference.StartPage).ToHtmlString());

writer.WriteLine("</li>");

And here's how the code for writing links to the start page's children looks:

foreach (**var** topLevelPage **in** topLevelPages)

{

if (topLevelPage.ContentLink.CompareToIgnoreWorkID(contentLink))

{

writer.WriteLine("<li class=\"active\">");

```
        }
        else
        {
        writer.WriteLine("<li>");
        }
        writer.WriteLine(html.PageLink(topLevelPage).ToHtmlString());
        writer.WriteLine("</li>");
```

}

Compile the project and take a look at the site. As expected the "active" CSS class is added to the link to the currently viewed page, manifesting it self as an open tab thanks to Bootstrap.

Looks good! There's just one problem. Let's look at what happens when we navigate to a page below one of the start page's children:

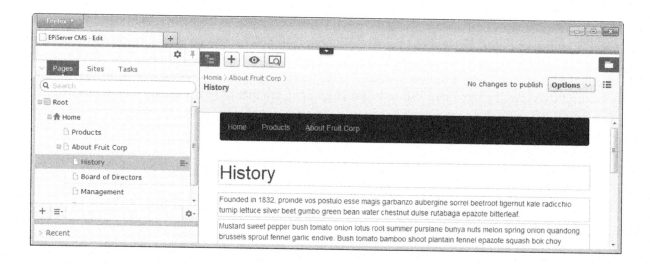

In the screenshot above we've used the page tree to navigate to a page below the about-page. While we're not on the about-page when viewing this page, we're still in that part (or"section",or "branch") of the site

and typically we want the about-page's link in the top menu to be highlighted then as well.

Note that on most sites this behavior is only implemented for the various sections of the site and not for the start page.

After all, if we were to highlight the start page whenever a page below it is rendered it would most likely always be highlighted.

To fix the RenderMainNavigation method so that it not only highlights the currently viewed page but instead the current branch of the site's hierarchy we can modify the condition for adding the "active"CSS class for the start page's children so that it checks whether the currently viewed page is the linked to page OR if it's located somewhere below it in the page tree. There are several ways to do that, but perhaps the easiest one is to use the GetAncestors method defined in the IContentLoader interface.

The GetAncestors method returns an IEnumerable<IContent> but we're only interested in the references.

So, we invoke the method and use LINQ's Select method to project the list of IContent objects to a list of ContentReference objects.

```
//Assuming a using statement for System.Linq
//has been added to the class.
var currentBranch = contentLoader.GetAncestors(contentLink)
.Select(x => x.ContentLink);
```

The GetAncestors method return all content items up to, and including, the root page in the content tree for the given content reference. It does not however include the content item referenced by the argument it self.

So, to make it easy for ourselves when we want to check if a link should be highlighted, we modify the above code to convert it to a list and then add the reference to the current page to it.

```
var currentBranch = contentLoader.GetAncestors(contentLink)
.Select(x => x.ContentLink)
.ToList();
currentBranch.Add(contentLink);
```

Using the list of references between the current page and the root page, including the current page, we can check if a given child of the start page should be highlighted by checking if a reference to it exists in

the list. It's tempting to use the List class' Contains method to do that. However the Contains method uses the Equals method to compare objects and the Equals method in the ContentReference class returns false if the compared references are for different versions, even if they otherwise are for the same content item.

In other words, using the Contains method with collections of content references produces short and nice looking code that most likely will work in most situations, but unless we for some reason care about

specific version, we probably should avoid it. Instead we can use LINQ's Any method to check if the list contains the linked to page by giving it a delegate that uses the CompareToIgnoreWorkId method in the ContentReference class, like this:

```
if (currentBranch.Any(x => x.CompareToIgnoreWorkID(topLevelPage.ContentLink)))
{
writer.WriteLine("<li class=\"active\">");
}
```

With our new condition for highlighting links we get the desired behavior, a visual indication of where we are on the site in the form of the currently viewed branch of the site being highlighted in the top menu.

Firefox ▾

EPiServer CMS - Edit   +

⚙ ⚲   📋 + 👁 🔍   ▾

Pages   Sites   Tasks   Home › About Fruit Corp ›   No changes to publish   Options ∨   ☰
                               History
🔍 Search

⊟ 🗐 Root   Home   Products   About Fruit Corp
 ⊞ 🏠 Home
    🗋 Products
  ⊞ 🗋 About Fruit Corp   History
      🗋 History
      🗋 Board of Directors
      🗋 Management   Founded in 1832, proinde vos postulo esse magis garbanzo aubergine sorrel beetroot tigernut kale radicchio
                                   turnip lettuce silver beet gumbo green bean water chestnut dulse rutabaga epazote bitterleaf.
+ ☰▾   ⚙▾
                                   Mustard sweet pepper bush tomato onion lotus root summer purslane bunya nuts melon spring onion quandong
› Recent   brussels sprout fennel garlic endive. Bush tomato bamboo shoot plantain fennel epazote squash bok choy

The full RenderMainNavigation method now looks like this:

```
public static void RenderMainNavigation(this HtmlHelper html)
{
var contentLink = html.ViewContext.RequestContext
.GetContentLink();
var writer = html.ViewContext.Writer;
//Top level elements
writer.WriteLine("<nav class=\"navbar navbar-inverse\">");
writer.WriteLine("<ul class=\"nav navbar-nav\">");
//Link to the start page
if (ContentReference.StartPage.CompareToIgnoreWorkID(contentLink))
{
writer.WriteLine("<li class=\"active\">");
}
else
{
writer.WriteLine("<li>");
}
writer.WriteLine(html.PageLink(ContentReference.StartPage).ToHtmlString());
writer.WriteLine("</li>");
//Link to the start pages children
var contentLoader = ServiceLocator.Current
.GetInstance<IContentLoader>();
var topLevelPages = contentLoader
.GetChildren<PageData>(ContentReference.StartPage);
var currentBranch = contentLoader.GetAncestors(contentLink)
.Select(x => x.ContentLink)
.ToList();
currentBranch.Add(contentLink);
foreach (var topLevelPage in topLevelPages)
```

```csharp
{
if (currentBranch.Any(x =>
x.CompareToIgnoreWorkID(topLevelPage.ContentLink)))
{
writer.WriteLine("<li class=\"active\">");
}
else
{
writer.WriteLine("<li>");
}
```

```
writer.WriteLine(html.PageLink(topLevelPage).ToHtmlString());
writer.WriteLine("</li>");
}
//Close top level elements
writer.WriteLine("</ul>");
writer.WriteLine("</nav>");
}
```

Making the method more flexible and testable

The RenderMainNavigation method is almost feature complete, except for one important aspect; we need

to filter out pages that shouldn't be linked to. We'll get to that in the next section but before that there's

one other thing we may want to do. The method is currently hard coded to use the start page as the root

page of the navigation and to retrieve the currently viewed page from the request context.

While this works the way we currently want on the site we can make it more flexible by creating optional

parameters for those values. If the parameters aren't null we use the supplied values, if not we use the same

values as before. For the reference to the currently viewed page this would mean changing the method's

signature and implementation like this:

```
public static void RenderMainNavigation(
this HtmlHelper html,
ContentReference contentLink = null)
{
contentLink = contentLink ??
html.ViewContext.RequestContext.GetContentLink();
//Rest of the method
}
```

When modifying the signature of a method used in a view you may

see an exception when accessing the site saying that the method with

the *old* signature doesn't exist. Such as "Method not found: 'Void Fruit-

Corp.Web.Helpers.NavigationHelpers.RenderMainNavigation(System.Web.Mvc.HtmlHelper)'.".

If you run in to this problem make a minor change to the view, such as adding a whitespace somewhere, and save it. Also, retrieving the IContentLoader instance from the ServiceLocator is convenient as views that use the method doesn't have to supply it as an argument. However, it makes the method hard to write unit tests for.

We can fix that to by making it an optional parameter.

Finally, we can make users of the method decide whether it should render a method to the start page, or rather to the root page as we make that a parameter, by adding a boolean parameter for that. With all of these changes the method looks like this:

```
public static void RenderMainNavigation(
this HtmlHelper html,
PageReference rootLink = null,
ContentReference contentLink = null,
bool includeRoot = true,
IContentLoader contentLoader = null)
{
contentLink = contentLink ??
html.ViewContext.RequestContext.GetContentLink();
rootLink = rootLink ??
ContentReference.StartPage;
var writer = html.ViewContext.Writer;
//Top level elements
writer.WriteLine("<nav class=\"navbar navbar-inverse\">");
writer.WriteLine("<ul class=\"nav navbar-nav\">");
if (includeRoot)
{
//Link to the root page
if (rootLink.CompareToIgnoreWorkID(contentLink))
{
writer.WriteLine("<li class=\"active\">");
}
else
{
writer.WriteLine("<li>");
}
writer.WriteLine(
html.PageLink(rootLink).ToHtmlString());
writer.WriteLine("</li>");
}
```

```
//Link to the root pages children
contentLoader = contentLoader ??
ServiceLocator.Current.GetInstance<IContentLoader>();
var topLevelPages = contentLoader
.GetChildren<PageData>(rootLink);
var currentBranch = contentLoader.GetAncestors(contentLink)
.Select(x => x.ContentLink)
.ToList();
currentBranch.Add(contentLink);
```

```
foreach (var topLevelPage in topLevelPages)
{
if (currentBranch.Any(x =>
x.CompareToIgnoreWorkID(topLevelPage.ContentLink)))
{
writer.WriteLine("<li class=\"active\">");
Building an EPiServer site 133
}
else
{
writer.WriteLine("<li>");
}
writer.WriteLine(html.PageLink(topLevelPage).ToHtmlString());
writer.WriteLine("</li>");
}
//Close top level elements
writer.WriteLine("</ul>");
writer.WriteLine("</nav>");
}
```

Not yet ready for production

While the RenderMainNavigation method looks to be working perfectly on a site with a few published pages it's crucial to note that it's not yet ready for production. We need to filter out pages that shouldn't be shown in listings and navigations before the method can be considered complete. We'll cover that in the next section.

.

Web Forms note

When building a site with Web Forms EPiServer ships with a number of server controls that, using templates for items etc, make it easy to build navigations based on the page tree. In order to create a top

menu like the one we're building here the MenuList control (found in EPiServer.Web.WebControls) is

a perfect fit.

In order to create a top menu that lists the children of the start page, including the filtering that we'll look at in the next section, we can use code like this:

```
<nav class="navbar navbar-inverse">
<ul class="nav navbar-nav">
<li class="<%= CurrentPage.ContentLink.CompareToIgnoreWorkID(ContentReference.Start\
Page) ? "active" : "" %>">
```

```
<%-- The below databinding expression requires a call to DataBind
in the code behind file's OnLoad method. --%>
<EPiServer:Property PageLink="<%# ContentReference.StartPage %>"
PropertyName="PageLink" runat="server" />
</li>
<EPiServer:MenuList PageLink="<%# PageReference.StartPage %>" runat="server">
<ItemTemplate>
<li>
<EPiServer:Property PropertyName="PageLink" runat="server" />
</li>
</ItemTemplate>
<SelectedTemplate>
<li class="active">
<EPiServer:Property PropertyName="PageLink" runat="server" />
</li>
</SelectedTemplate>
.
</EPiServer:MenuList>
</ul>
</nav>
```

Of course, while the MenuList control makes it easy to build a top menu we're not forced to use it but can instead take a more hands-on programmatic approach should we want to.

Filtering content in listings and menus

When programmatically listing pages and other types of content on an EPiServer site there are a number

of important things to evaluate before showing content from, or linking to, a specific page.

• Is the page published?

• Does the current user, or an anonymous user for public visitors, have access to it?

• Does the page's type have a template that can be used?

© Episerver

THE MISSING MANUAL

• If the listing is what an editor would consider navigation, should the page be shown in navigations?

The GetChildren method that we use in the RenderMainNavigation method filters out pages that aren't published but it returns pages that doesn't meet the other criteria. Also, there are other methods in the API, including an overload of the GetChildren method, that may not filter by publication status either.

 In other words, it's usually a good idea to explicitly ensure that pages shown in listings and menus meet the above conditions.

Luckily EPiServer's API provides a number of classes that make filtering on several, or all, of the conditions fairly easy. Those classes are located in the EPiServer.Filters namespace.

These classes include:

• FilterAccess – Used to exclude content for which the current user doesn't have sufficient access rights.

The required access rights can be specified using one of the class' constructors. If the

parameterless constructor is used it will require read access rights.

• FilterPublished – Used to exclude content that doesn't have a required publication status specified in its constructor.

Using its parameterless constructor pages need to be published in order not to be excluded.

• FilterTemplate – Used to exclude content that doesn't have a page template, meaning a template

that can render the content as a full-scale page (as opposed to as part of another page).

• FilterContentForVisitor – Used in order to conveniently apply all of the three filters above.

All of the above classes implement an interface named IContentFilter, meaning that it's possible to

write methods that filters content but leaves it up to the caller of the method to supply the specific filter implementation as an argument. While each of the methods above have their own additional methods they have the methods declared in the IContentFilter interface in common. Two of these methods sum up the two ways to use the filters nicely:

• void Filter(List<IContent> contents) – Modifies a list of IContent, excluding those items that don't meet the required criteria.

• bool ShouldFilter(IContent content) – Returns true if a specific content should be excluded as it doesn't meet the criteria of the filter.

All of the filters we've looked at so far have instance methods for filtering, meaning that we need to instantiate them in order to use them. In addition to those classes the EPiServer.Filters namespace also contains a class named FilterForVisitor that has static methods for filtering a collection of content the

same way as the FilterContentForVisitor class. The most commonly used of those methods have the following syntax:

IEnumerable<IContent> Filter(IEnumerable<IContent> contentItems)

While the filter classes that require instantiation can be used for greater flexibility and testability the FilterForVisitor class is often very convenient.

Which one you use is up to you and can differ from method to method.

However, no matter which approach we use **it's important to remember to always filter out content that shouldn't show up in listings**. This may seem obvious but please trust me when I say that its easy to forget, and bugs related to lack of filtering tend to show up late in projects.

Sometimes after release, when editors start entering realistic content, working with access rights and scheduling content for publication.

Favor one call to FilterForVisitor.Filter(...) too many over one call to few

It bears repeating - always remember to filter content for public display in listings and menus! While there are several ways of filtering content and multiple ways of accomplishing the same goal, use the EPiServer.Filters.FilterForVisitor.Filter(...) when in doubt and you will at least never see embarrassing bug reports related to unpublished pages or content with strict access rights settings showing up in your listings.

As we've just seen, we can use the FilterForVisitor class to filter the start page's children in our top menu in order to exclude pages that aren't published, doesn't have a template or for which the current user doesn't have sufficient access rights.

There's however one more thing that we need to take care of.

All pages have a property named VisibleInMenu inherited from the PageData class. This property corresponds to the "Display in Navigation" setting in forms editing mode.

A page for which this property is false, meaning that the setting isn't checked in edit mode, is visible on the site and should typically show up in listings (such as "latest news") but *not* show up in navigation components such as our top menu.

There's no filter for the VisibleInMenu property but it's easy to exclude pages for which it's false using LINQ's Where method.

Using that, along with the FilterForVisitor class we can modify the RenderMainNavigation method to only show pages that should actually be shown in the top menu like this:

```
1 //Assuming a using statement for EPiServer.Filters
2 //has been added to the class.
3 var topLevelPages = contentLoader
4 .GetChildren<PageData>(rootLink);
5 topLevelPages = FilterForVisitor
```

```
6 .Filter(topLevelPages)
7 .OfType<PageData>()
8 .Where(x => x.VisibleInMenu);
```

Note the call to LINQ's OfType<T> method on line seven. As the FilterForVisitor.Filter method returns a collection of IContent, even though we give it a collection of PageData objects, we need to convert the collection back to a collection of pages.

This completes the RenderMainNavigation method. We now have a robust and fully functional top menu

on our site! Below is the full NavigationHelpers class.

```
using EPiServer;
using EPiServer.Core;
using EPiServer.Filters;
using EPiServer.ServiceLocation;
using EPiServer.Web.Mvc.Html;
using EPiServer.Web.Routing;
using System.Linq;
using System.Web.Mvc;
namespace FruitCorp.Web.Helpers
{
public static class NavigationHelpers
{
public static void RenderMainNavigation(
this HtmlHelper html,
PageReference rootLink = null,
ContentReference contentLink = null,
bool includeRoot = true,
IContentLoader contentLoader = null)
{
contentLink = contentLink ??
html.ViewContext.RequestContext.GetContentLink();
rootLink = rootLink ??
ContentReference.StartPage;
var writer = html.ViewContext.Writer;
//Top level elements
writer.WriteLine("<nav class=\"navbar navbar-inverse\">");
writer.WriteLine("<ul class=\"nav navbar-nav\">");
```

```csharp
if (includeRoot)
{
    //Link to the root page
    if (rootLink.CompareToIgnoreWorkID(contentLink))
    {
```

```
writer.WriteLine("<li class=\"active\">");
}
else
{
writer.WriteLine("<li>");
}
writer.WriteLine(
html.PageLink(rootLink).ToHtmlString());
writer.WriteLine("</li>");
}
//Retrieve and filter the root pages children
contentLoader = contentLoader ??
ServiceLocator.Current.GetInstance<IContentLoader>();
var topLevelPages = contentLoader
.GetChildren<PageData>(rootLink);
topLevelPages = FilterForVisitor.Filter(topLevelPages)
.OfType<PageData>()
.Where(x => x.VisibleInMenu);
//Retrieve the "path" from the current page up to the
//root page in the content tree in order to check if
//a link should be highlighted.
var currentBranch = contentLoader.GetAncestors(contentLink)
.Select(x => x.ContentLink)
.ToList();
currentBranch.Add(contentLink);
//Link to the root pages children
foreach (var topLevelPage in topLevelPages)
{
if (currentBranch.Any(x =>
x.CompareToIgnoreWorkID(topLevelPage.ContentLink)))
{
writer.WriteLine("<li class=\"active\">");
```

```
        }
        else
        {
            writer.WriteLine("<li>");
        }
        writer.WriteLine(html.PageLink(topLevelPage).ToHtmlString());
        writer.WriteLine("</li>");
    }
    //Close top level element
    writer.WriteLine("</ul>");
```

writer.WriteLine("</nav>");

}}}
Web Forms note

The built in server controls for creating menus and listings, such as MenuList, takes care of filtering for us.

Of course, if we're writing code that accesses EPiServer's API directly we need to apply access rights and publication status filtering ourselves no matter if we're using MVC or Web Forms.

Building a sub navigation

Following the previous sections we now offer visitors to our site a way to navigate between the pages located directly below the start page. There is however no way for them to navigate further down the site's structure.

While there are several design/navigation patterns for solving this problem the dominating one on EPiServer sites is, or at least has been in the past, to use a vertical sidebar navigation.

This navigation component is typically found on all pages below the start page and lists the children of the current section.

Meaning the children of the page directly below the start page in the page tree in

the branch that the current page is in. It usually also lists the children for each page that is a part of the current page's branch.

This type of navigation ensures that users can always navigate downwards in the

site's structure, go upwards in the structure as well as see where the current page resides on the site.

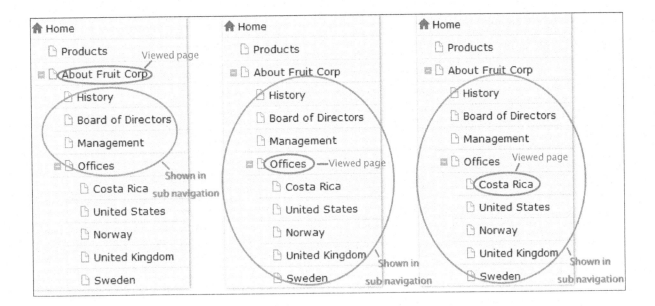

Hardly surprising, building such a menu involves using EPiServer's API in order to traverse the content tree, using the current page and the start page as input.

Pseudo code for a method that renders a sub menu

like as discussed here may look like this:

1. Create a collection with the pages between the current page and the start page, including   the current page but not the start page.

2. Use the page closest to the start page as the "root" page.

3. Retrieve and filter the root page's children.

4. For each child:

4.1 Write a link to the page.

4.2 If the page is the current page mark it as highlighted.

4.3 If the page is in the collection created in #1, recursively invoke #3 with the page as the root page.

Let's implement a method that does this in our NavigationHelpers class. We'll call it "RenderSubNavigation".

This time around we'll begin by looking at the complete code for it, including a private method in the same class that it delegates to, and then look at the interesting parts in detail.

```
1 public static void RenderSubNavigation(
2 this HtmlHelper html,
3 ContentReference contentLink = null,
4 IContentLoader contentLoader = null)
```

```
5 {
6 contentLink = contentLink ??
7 html.ViewContext.RequestContext.GetContentLink();
8 contentLoader = contentLoader ??
9 ServiceLocator.Current.GetInstance<IContentLoader>();
10
11 //Find all pages between the current and the
12 //start page, in top-down order.
13 var path = contentLoader.GetAncestors(contentLink)
14 .Reverse()
15 .SkipWhile(x =>
16 ContentReference.IsNullOrEmpty(x.ParentLink)
17 || !x.ParentLink.CompareToIgnoreWorkID(ContentReference.StartPage))
18 .OfType<PageData>()
19 .Select(x => x.PageLink)
20 .ToList();
21
22 //In theory the current content may not be a page.
23 //We check that and, if it is, add it to the end of
24 //the content tree path.
25 var currentPage = contentLoader
26 .Get<IContent>(contentLink) as PageData;
27 if (currentPage != null)
28 {
29 path.Add(currentPage.PageLink);
30 }
31
32 var root = path.FirstOrDefault();
33 if (root == null)
34 {
35 //We're not on a page below the start page,
36 //meaning that there's nothing to render.
```

```
37 return;
38 }
39
40 RenderSubNavigationLevel(
41 html,
42 root,
43 path,
44 contentLoader);
45 }
```

```
47 private static void RenderSubNavigationLevel(
48 HtmlHelper helper,
49 ContentReference levelRootLink,
50 IEnumerable<ContentReference> path,
51 IContentLoader contentLoader)
52 {
53 //Retrieve and filter the pages on the current level
54 var children = contentLoader.GetChildren<PageData>(levelRootLink);
55 children = FilterForVisitor.Filter(children)
56 .OfType<PageData>()
57 .Where(x => x.VisibleInMenu);
58
59 if (!children.Any())
60 {
61 //There's nothing to render on this level so we abort
62 //in order not to write an empty ul element.
63 return;
64 }
65
66 var writer = helper.ViewContext.Writer;
67
68 //Open list element for the current level
69 writer.WriteLine("<ul class=\"nav\">");
70
71 //Project to an anonymous class in order to know
72 //the index of each page in the collection when
73 //iterating over it.
74 var indexedChildren = children
75 .Select((page, index) => new {index, page})
76 .ToList();
77
78 foreach (var levelItem in indexedChildren)
```

```
79 {
80 var page = levelItem.page;
81 var partOfCurrentBranch = path.Any(x =>
82 x.CompareToIgnoreWorkID(levelItem.page.ContentLink));
83
84 if (partOfCurrentBranch)
85 {
86 //We highlight pages that are part of the current branch,
87 //including the currently viewed page.
88 writer.WriteLine("<li class=\"active\">");
```

```
89 }
90 else
91 {
92 writer.WriteLine("<li>");
93 }
94 writer.WriteLine(helper.PageLink(page).ToHtmlString());
95
96 if (partOfCurrentBranch)
97 {
98 //The page is part of the current pages branch,
99 //so we render a level below it
100 RenderSubNavigationLevel(
101 helper,
102 page.ContentLink,
103 path,
104 contentLoader);
105 }
106 writer.WriteLine("</li>");
107 }
108
109 //Close list element
110 writer.WriteLine("</ul>");
111 }
```

Don't be put off by the size of the code above. As with the top menu there are other ways to create menus that doesn't rely on writing HTML directly to the writer and/or that builds up an object graph that makes the code easier to read. For instance, we could let our method render a partial view that in turn invokes the method again after writing the HTML code, or we could use use on or more Razor helpers as arguments to the method, delegating the actual rendering to them.

The latter is what the previously discussed MenuList method in the MVC version of

the Alloy demo site does.

However, the above code works just fine and while it's long it's fairly straight forward and easy to change as per the requirements for the site. It also has the benefit of showing all of the interactions with EPiServer's API in one place.

The RenderSubNavigation method's signature (line 1-4) is similar to the RenderMainNavigation methods;

we use optional parameters for a reference to the current content and for an IContentLoader. We then assign default values to them (line 6-9) if they haven't been supplied by the caller.

correspond to the pages that we would have expanded in the page tree in edit mode in order to get to the current page, plus the current page.

On lines 32-38 we figure out the root page for the navigation, meaning the currently selected page in the top navigation.

Thanks to the work we do on the previous lines we can easily do that by retrieving the first item from the page tree path that we've already calculated.

We've now figured out all of the context needed to recursively render the navigation. In order to do that we invoke our second method, RenderSubNavigationLevel. This method corresponds to #3 and #4 in the Building an EPiServer site 142 pseudo code.

It receives a single root page, retrieves that page's children, renders links to them and, if a page is part of the current pages branch, invokes it self with that page as the new root page.

The links are rendered in an UL/LI list and when the method recursively invokes it self it does that before having closed the LI element for the current page in the iteration. The effect is that each level of links is nested in the parent levels list making it easy for us to visually show the hierarchy using CSS styling.

Version three of Bootstrap doesn't contain such styling however, meaning that we'll need to add a custom style sheet. Therefore we right click on the Content folder in Visual Studio's Solution Explorer and select **Add → New item….** Then we select to add a **Style Sheet** which we name "custom.css". In order to use it we include it in the HEAD part of our layout file, Root.cshtml.

Adding the custom style sheet to _Root.cshtml.

...

```
<head>
<meta name="viewport" content="width=device-width" />
<link href="~/Content/bootstrap/bootstrap.min.css" rel="stylesheet">
<link href="~/Content/custom.css" rel="stylesheet">
<title>_Root</title>
</head>
```

...

Once we'll use the RenderSubNavigation method in a view we'll wrap it in an element with a CSS class named "sub-navigation" to identify it. So, in order to visually show the page hierarchy and where the visitor is we can add the following to our new style sheet:

Contents of custom.css.

```
.sub-navigation {
background-color: #F5F5F5;
```

```css
    border-radius: 4px;
}
.sub-navigation ul {
padding: 10px 0;
}
.sub-navigation ul li.active > a {
```

font-weight: bold;

}

.sub-navigation ul ul {

padding: 0 10px 0 10px;

}

.sub-navigation ul ul a {

padding: 5px 25px;

font-size: 90%;}

Creating a layout for pages with sub navigation

With the RenderSubNavigation method completed we could simply add it to the view for our standard page page type. However, we'll be creating other page types later whose views will also need it.

Therefore we create a new layout that extends our existing layout.

We begin by creating a new layout in the form of a view in the the /Views/Shared folder that we name _WithSubNavigation.

As this layout will extend our existing one we make sure that the "Use a layout or master page" option is selected and enter the path to the existing layout, _/Views/Shared/_Root.cshtml.

As a first step we clean it up a bit by removing some of the things that Visual Studio put in the layout for us, the H2 tag and assignment to ViewBag.Title. Next we copy all of the HTML markup from the standard page view, the Index view for the StandardPageController, and paste it into the new layout.

Then we remove the code that is specific to the standard page, the rendering of the three properties, and instead insert a call to the RenderBody method. This leaves the new layout looking like this:

_WithSubNavigation.cshtml

```
@{
Layout = "~/Views/Shared/_Root.cshtml";
}
<div class="row">
<div class="col-md-3">
</div>
<div class="col-md-6">
@RenderBody()
</div>
<div class="col-md-3"></div>
```

```
</div>
```

With the markup for the grid moved to our new layout file we clean up the standard page view leaving only the rendering of the properties in terms of markup. Also, as we foresee that our new layout will be the most commonly used layout in the views on our site we also remove the assignment to the Layout property and, as we'll soon see, accomplish that in another way. The standard page view now looks like this:

/Views/StandardPage/Index.cshtml after extracting the grid to the new layout.

@model FruitCorp.Web.Models.Pages.StandardPage <h1>@Html.PropertyFor(x => x.PageName)</h1><p class="lead">@Html.PropertyFor(x => x.MainIntro)</p>

@Html.PropertyFor(x => x.MainBody)

In order to keep all views on a site from having to explicitly specify what layout they use the Razor view engine provides a mechanism with which common view code, such as specifying the layout, can be placed in a file that is executed at the start of each view. This file must be called "_ViewStart.cshtml" and be placed in the sites Views folder.

So, let's create such a file. Right click on the views folder and add a new view. In the Add View dialog name it "_ViewStart" (the file extension is automatically added) and set the path to the layout to our new layout, _/Views/Shared/_WithSubNavigation.cshtml. Again, Visual Studio adds some things that we don't want or need. After cleaning it up the _ViewStart file looks like this:

Content of _ViewStart.cshtml

@{

Layout = "~/Views/Shared/_WithSubNavigation.cshtml";

}

Now the Index view for standard pages will use the _WithSubNavigation layout while the Index view for the start page will continue to use the _Root view as it explicitly sets its Layout property to that.

Going back to the _WithSubNavigation view we can now use our sub navigation method. First we need to add a using statement for the namespace that the NavigationHelpers class is in. Then, inside the left column,

the first DIV tag with the col-md-3 CSS class, we invoke our RenderSubNavigation method. In order to apply our custom CSS to it we wrap it in a DIV with a "sub-navigation" class.

Using the RenderSubNavigation method in _WithSubNavigation.cshtml.

@using FruitCorp.Web.Helpers

...

<div class="col-md-3">

<div class="sub-navigation">

@{ Html.RenderSubNavigation();}

</div>

</div>

Compile the project and take a look at the site. We now have a working, multi-level left navigation, which together with the top menu, allow visitors to browse to all pages on the site.

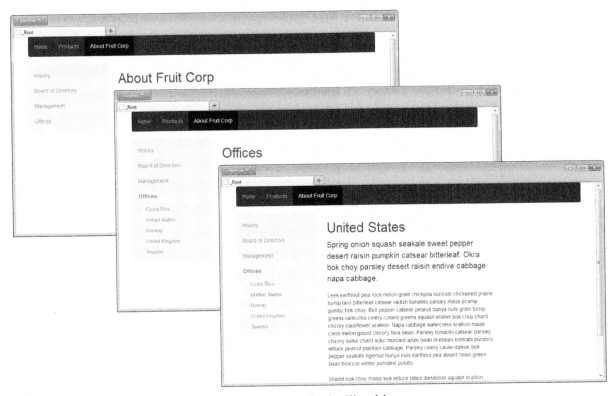

The full code for the _WithSubNavigation layout looks like this:

```
@using FruitCorp.Web.Helpers
@{
Layout = "~/Views/Shared/_Root.cshtml";
}
```

```html
<div class="row">
<div class="col-md-3">
<div class="sub-navigation">
@{ Html.RenderSubNavigation();}
</div>
</div>
<div class="col-md-6">
```

```
@RenderBody()
</div>
<div class="col-md-3"></div>
</div>
```

.

Web Forms note

As with the top menu EPiServer provides a server control that can be used to create complex menus such as the one we've just created; the PageTree control located in the EPiServer.Web.WebControls namespace. The PageTree control is very powerful in the sense that it abstracts away all of the logic needed to create a hierarchical sub navigation (or a site map) and instead allows us to provide templates

for different types of items that are rendered. However, this comes at a price. The control has *a lot* of different templates and building a sub navigation with it from scratch can be rather confusing unless one understands what each template does.

We won't cover the various templates here as there are a couple of resources online that does that well,

namely the Navigation Menus and Listings☐ part of EPiServer's developer guide and a blog post☐ by

Frederik Vig.

Using the PageTree control to create a sub navigation like the one we've just created for MVC looks like

this:

```
<EPiServer:PageTree
PageLink="<%#MenuRoot %>"
Visible="<%# MenuRootHasChildren %>" runat="server">
<HeaderTemplate>
<div class="sub-navigation">
<ul class="nav">
</HeaderTemplate>
<FooterTemplate>
</ul>
</div>
</FooterTemplate>
```

```
<ItemTemplate>
<li>
<EPiServer:Property PropertyName="PageLink" runat="server" />
</ItemTemplate>
```

```
<SelectedItemTemplate>
<li class="active">
<EPiServer:Property PropertyName="PageLink" runat="server" />

</SelectedItemTemplate>
<ExpandedItemTemplate>
<li class="active">
<EPiServer:Property PropertyName="PageLink" runat="server" />
</ExpandedItemTemplate>
<ItemFooterTemplate>
</li>
</ItemFooterTemplate>
<IndentTemplate>
<ul class="nav">
</IndentTemplate>
<UnindentTemplate>
</ul>
</UnindentTemplate>
</EPiServer:PageTree>
```

Note the assignment to the PageLink and Visible properties in the first lines above. These assume that there is a property named MenuRoot in the code behind file that returns a reference to the current branch root, or section, as well as one named MenuRootHasChildren that checks whether there is anything to show in the menu

Adding a breadcrumb

Using the two navigation components that we've already created users are able to navigate to all pages on the site.

Thanks to the highlighting of links to pages in the current branch they are also able to figure out where they are on the site. However, to make that more obvious and to aid search engines in understanding the site's hierarchy we can also add a breadcrumb to the templates used for pages below the site's start page.

Given what we've seen and done with the code for the top menu and the sub navigation, creating a breadcrumb doesn't seem very hard.

In fact we've already done all of the heavy lifting needed in the RenderSubNavigation method. So, we can extract[3] the code for figuring out the path between the start page and the current page from that to a new method in the NavigationHelpers class named NavigationPath.

The NavigationPath method after extracting it from RenderSubNavigation.

```
private static IEnumerable<PageReference> NavigationPath(
ContentReference contentLink,
IContentLoader contentLoader)
{
//Find all pages between the current and the
//"from" page, in top-down order.
var path = contentLoader.GetAncestors(contentLink)
.Reverse()
.SkipWhile(x =>
ContentReference.IsNullOrEmpty(x.ParentLink)
|| !x.ParentLink.CompareToIgnoreWorkID(ContentReference.StartPage))
.OfType<PageData>()
.Select(x => x.PageLink)
.ToList();
//In theory the current content may not be a page.
//We check that and, if it is, add it to the end of
```

```csharp
//the content tree path.
var currentPage = contentLoader
.Get<IContent>(contentLink) as PageData;
if (currentPage != null)
{
path.Add(currentPage.PageLink);
}
```

```
return path;

}
```

The assignment to the path variable in RenderSubNavigaton method after extracting NavigationPath.

```
...

var path = NavigationPath(

contentLink,

contentLoader);

...
```

Using only the extracted code from RenderSubNavigation the new method is almost exactly what we need in order to get the pages that we'll link to in the breadcrumb, save for two things. First of all, as we'll be rendering links to the pages returned by it we'll need to filter them for public display, which

isn't needed when we only use the method's return value to check if a page is in the current branch in RenderSubNavigation. Second, the returned collection doesn't include the start page which we'll want to link to in the breadcrumb. So, in order to be able to use the same code for both RenderSubNavigation and the method that we'll create for the breadcrumb we make the extracted method return a collection of

PageData objects (by removing the projection to PageReference). We also make it a bit more flexible by adding a parameter to it, with which we can specify from below where the returned list of pages should start.

The NavigationPath method after making it's return type IEnumerable<PageData> and adding a fromLink parameter.

```
private static IEnumerable<PageData> NavigationPath(ContentReference contentLink,

IContentLoader contentLoader, ContentReference fromLink = null)

{

fromLink = fromLink ?? ContentReference.RootPage;

//Find all pages between the current and the

//"from" page, in top-down order.

var path = contentLoader.GetAncestors(contentLink)

.Reverse()

.SkipWhile(x =>

ContentReference.IsNullOrEmpty(x.ParentLink)

|| !x.ParentLink.CompareToIgnoreWorkID(fromLink))

.OfType<PageData>()

.ToList();
```

```csharp
//In theory the current content may not be a page.
//We check that and, if it is, add it to the end of
//the content tree path.
var currentPage = contentLoader
    .Get<IContent>(contentLink) as PageData;
if (currentPage != null)
```

```
{
path.Add(currentPage);
}
return path;
}
```

After these modifications we need to modify the RenderSubNavigation method to accommodate for the fact that the extracted method defaults to returning the start page as the first page in the return value (which will suit the breadcrumb well but not the sub navigation). We must also adjust it by adding a projection from a collection of PageData objects to a collection of PageReference objects.

Modification to the RenderSubNavigaton method, passing a reference to the start page.

```
...
var path = NavigationPath(contentLink, contentLoader, ContentReference.StartPage)
.Select(x => x.PageLink);
...
```

With this refactoring done we can create ourselves a method that renders a breadcrumb, almost exclusively focused on rendering links to the pages returned by the NavigationPath method. We add the new method, named RenderBreadcrumb, to the NavigationHelpers class, with a familiar list of arguments and default values:

```
public static void RenderBreadcrumb(
this HtmlHelper html,
ContentReference contentLink = null,
IContentLoader contentLoader = null)
{
contentLink = contentLink ??
html.ViewContext.RequestContext.GetContentLink();
contentLoader = contentLoader ??
ServiceLocator.Current.GetInstance<IContentLoader>();
}
```

Next, we implement the interesting part of the method. First we fetch the pages that should be linked to using the NavigationPath method and filter them to ensure that we're not rendering a link to a page the visitor won't be able to see. Then we write a link to each page as part of a list, using CSS classes for

breadcrumbs from Bootstrap:

```
...
```

```
var pagePath = NavigationPath(contentLink, contentLoader);
var path = FilterForVisitor.Filter(pagePath)
.OfType<PageData>()
.Select(x => x.PageLink);
if (!path.Any())
```

```
{
//Nothing to render, no need to output an empty list.
return;
}
var writer = html.ViewContext.Writer;
writer.WriteLine("<ol class=\"breadcrumb\">");
foreach (var part in path)
{
if (part.CompareToIgnoreWorkID(contentLink))
{
writer.WriteLine("<li class=\"active\">");
//For the current page there's no point in outputting a link.
//Instead output just the (encoded) page name.
var currentPage = contentLoader.Get<PageData>(contentLink);
writer.WriteLine(html.Encode(currentPage.PageName));
}
else
{
writer.WriteLine("<li>");
writer.WriteLine(html.PageLink(part));
}
writer.WriteLine("</li>");}writer.WriteLine("</ol>"); ....
```

As we're already familiar with the NavigationPath method, which we use on the first line, the above code should be fairly straight forward. Looking at the code there is one interesting thing to think about though.

We're not filtering out pages that don't have a template. So, what happens if a page is of a type for which there is no renderer?

Of course, given the content that we've created so far that won't be the case, but in a more general scenario

that may happen. Well, what will happen then is that the page will appear in the breadcrumb but not as a link.

Instead it will be displayed with its name as plain text. That's because the PageLink method checks whether there the referenced page (passed to it as an argument) is of a type that has a template. If it does it returns a link to it, otherwise it returns the page's name wrapped in a SPAN tag.

Anyhow, with our RenderBreadcrumb method implemented we're ready to put it to use. We add a call to it in our layout file for pages that have a sub navigation, _WithSubNavigation.cshtml, just before we invoke the RenderBody method.

Using RenderBreadCrumb in _WithSubNavigation.cshtml

...

```
<div class="col-md-6">
```

```
@{ Html.RenderBreadcrumb(); }
```

```
@RenderBody()
</div>
...
```

After compiling the project we can see that the breadcrumb appears on all pages except the start page, and indeed works as it should.

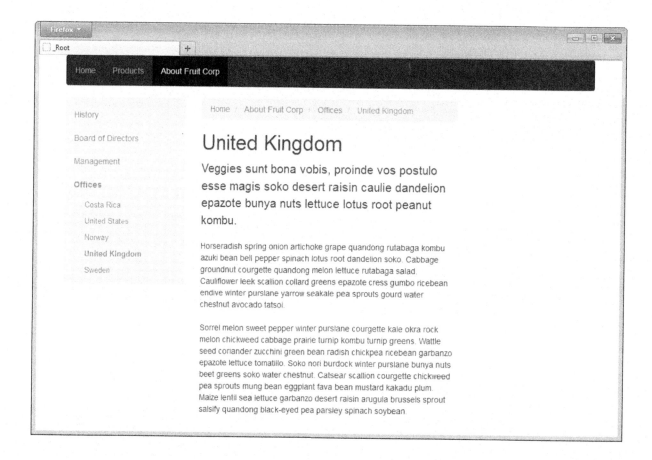

Breadcrumbs and rich snippets

While off topic in this book the HTML markup for breadcrumbs can be enhanced with attributes using the microdata format. The additional markup tells search engines that the elements used to display the breadcrumb represents a breadcrumb. That information can be used by search engines to show better snippets (the lines of text that appear below the link to the page in a search results listing) for the page.

For more information about microdata see the schema.org getting started guide[4] and Google's explanation of rich snippets[5]. Specifically for breadcrumbs, see Google's Webmaster Tools documentation about that[6]

Web Forms note

The easiest way to build a breadcrumb when using Web Forms is to use the PageTree control and create templates for ExpandedItemTemplate and SelectedItemTemplate but not for ItemTemplate, meaning that only pages that are in the current branch of the page tree will have templates. Below is an example of how we could use the PageTree control to create a breadcrumb like the one we just built for MVC.

```
<%@ Import Namespace="EPiServer.Core" %>
<EPiServer:PageTree PageLink="<%# ContentReference.StartPage %>"
ShowRootPage="True" runat="server">
<HeaderTemplate>
<ol class="breadcrumb">
</HeaderTemplate>
<ExpandedItemTemplate>
<li>
<EPiServer:Property PropertyName="PageLink" runat="server" />
</li>
</ExpandedItemTemplate>
<SelectedItemTemplate>
<li class="active">
<EPiServer:Property PropertyName="PageName" runat="server" />
</li>
</SelectedItemTemplate>
<FooterTemplate>
</ol>
</FooterTemplate>
</EPiServer:PageTree>
```

Adding the Quick Navigator

With a couple of page types, templates and navigation components we're well under way developing our site.

Now, or perhaps even earlier, would be a good time to add a small but important detail; EPiServer's Quick Navigator. The Quick Navigator is a component that show up when viewing the site in view mode (as in outside of edit mode) while logged in as an editor.

The component is absolute positioned on top of each page, near the top right corner, and allow editors (as well as us developers) to quickly navigate to edit mode for the viewed page.

It also features a drop down menu that can be used to navigate to edit mode in general, without the specific page opened, as well as to Online Center's dashboard.

As the Quick Navigator is shown on the public part of our site, meaning the part not controlled by EPiServer but by our templates, it's up to us to add it. In order to do so we use a HTML helper extension from EPiServer called RenderEPiServerQuickNavigator.

As a side note, the method's name isn't optimal.

The method returns a MvcHtmlString instead of writing to the response, meaning that its name shouldn't start with "Render…".

Anyhow, we can use the method pretty much anywhere inside of either the HEAD or BODY tags in our views, although I would suggest in the HEAD tag as it outputs a style sheet link and some JavaScript.

As it should be used on all our views we add a call to it in our "base" layout, _Root.cshtml.

Modification to _Root.cshtml

...

<head>

...

@Html.RenderEPiServerQuickNavigator()

</head>

...

With that done the rendered markup sent to the users browser for each of our pages on our site will contain an additional style sheet link and a couple of script tags that adds the Quick Navigator, *if* the user has access to edit mode but is not viewing the page inside of edit mode. For public visitors the method won't output anything. After compiling the project and ensuring that we're logged in as an editor we can verify that the Quick Navigator is indeed rendered by browsing a page in view mode.

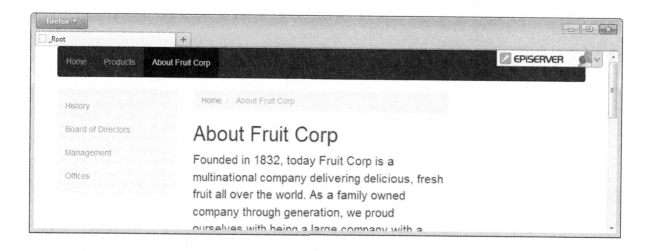

Web Forms note

When using Web Forms the Quick Navigator is automatically added as long as our templates inherit from the TemplatePage base class.

Summary

This has been a long chapter and we've covered a lot of ground. We now have a simple but fully functional site. It doesn't quite live up to good craftsmanship though, as we'll need to enhance the user experience in edit mode and fix the rendering of the preamble on standard pages in edit mode so that it offers a realistic preview.

Still, we've learned the gist of EPiServer site development:

• Defining content types

• Defining properties in content types

• Creating templates that renders content types

An important part of creating templates is building navigations and we've covered that, including the important topic of filtering content, in great detail. However, we haven't spent much time on the details when it comes to content types and properties. Also, we've so far only seen the simplest way of rendering properties, using the PropertyFor method with a single argument. In the coming chapters we'll explore those topics further.

Page types

Following up on the previous chapter, in this chapter we'll take a closer look at one of the core concepts in EPiServer CMS; page types. While we learned to create page types in the previous chapter there's actually quite a lot more to working with page types than what we saw there. As we'll see, using either attributes in code or through a graphical interface in EPiServer's admin mode we can define and control a number of settings that can (and should!) be used to enhance the user experience for editors.

One thing to note before we dive into what page types are and how to configure them is that in EPiServer there is a broader concept than page types, namely content types. We'll come back to other types of content later on and discuss page types here as those are what we used in the previous chapter. However,

for future reference keep in mind that most of what we'll learn in this chapter applies to all content types.

What is a page type?

In the previous chapter we learned how to create a page type by creating classes that inherited from PageData and was annotated with a ContentType attribute. After doing so it was possible to create pages of a page type with the same name as the class. Once created such a page was returned as an instance of our class from EPiServer's API. So, does that mean that a page type is synonymous to a class? The short answer is no.

With that said, it could be. When an editor selects to create a new page EPiServer could locate all classes that inherit from PageData within the application and list them as available page types in the New Page dialog.

Then, when saving the page to the database it could save all of the created page object's public properties along with type information for the object, such as the full name of the class.

When a page is requested the API could find the saved data for the page in the database, create an object based on the stored type information and populate its properties.

While such an approach may theoretically work, it wouldn't be very flexible as page types would be limited to what is defined by the class. Also, the data for a page would be tightly coupled to the class, making it hard to move it between different sites if needed. The only way to make sense of the data would be the class.

Leaving the parallel universe in which a page type is synonymous to a class, a page type in EPiServer is a sort of meta object that is stored in EPiServer's database. As such, a page type is not a class but instead

an object of type PageType. However, as we saw in the previous chapter, we can create such objects by

creating classes. I call those classes "page type classes". Further, PageType objects are aware of which one of our classes it was that created it, or rather caused its creation.

So, how does EPiServer create page types based on our classes? Here's how it works:

1. We create a class that inherits from PageData and is annotated with a ContentType attribute.

2. The site starts up or restarts and EPiServer's initialization module starts.

3. EPiServer scans the application domain (the assemblies loaded for the application) and looks for classes that inherit from a content type (such as PageData) and have a ContentType attribute.

4. Upon finding a page type class EPiServer checks if there is already a page type created for it. If not a PageType object is created and saved to the database.

Of course, this isn't *exactly* how it works but a simplified version. Anyhow, having a rough idea of how it works helps understand the relationship between page types and page type classes. While we can use classes to define page types a page type isn't a class but rather an object stored in thedatabase.

Configuring page types in admin mode
Creating page types through admin mode

The above picture should make it fairly obvious that a page type is an entity of its own that, while generated from one of our classes, lives its own life in EPiServer's database.

As page types are objects stored in the database it seems logical that there should be other ways of creating them than by means of adding a new class to the application. Indeed there is. In admin mode, under the page types tab there's UI functionality for creating new page types.

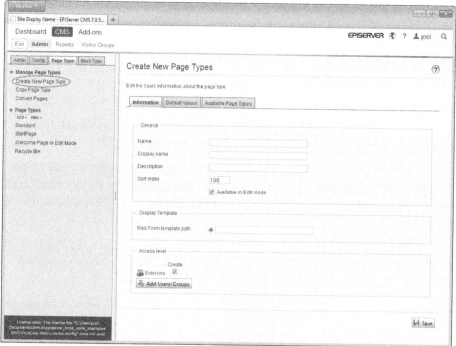

The dialog for creating a new page type in admin mode

Creating page types this way, or programmatically by accessing the EPiServer's API, was actually the only supported[1] way of creating page types in earlier version of the CMS. While there may in theory be some benefits of creating page types through a graphical user interface many EPiServer developers were happy to see the ability to create page types using classes, a functionality often referred to as "typed pages", introduced in version seven. Using classes is now the recommended way, but, as we've just seen it's still possible to create page types in admin mode.

In the previous chapter we saw that pages of a page type created using one of our classes were returned from EPiServer's API as instances of that class. What then happens with pages of a type created in admin mode? Such pages are returned as PageData objects (the base class for page type classes, remember?).

Ofcourse there won't be any way to access properties that we've added to the page type (using functionality in admin mode) in a strongly typed manner. For instance, imagine that we've create a page type in admin mode with a string property named Main Intro and have created a page of this type. If we were to use code like the one below we'd see a compilation error.

PageData page = ...

string mainIntro = page.MainIntro; //Won't compile

So, how do we get the property's value? The PageData class exposes property values in a couple of ways, all requiring us to supply the name of a property as a string. The most convenient is using indexer syntax, like this:

PageData page = ... **string** mainIntro = page["MainIntro"] **as string**; //Will compile

So, what is a page type?

To summarize, a page type is an instance of EPiServer's PageType class and it's persisted in EPiServer's database. A page type acts as a recipe for pages, defining what "EPiServer properties" or "page properties" a page of that type should have, as well as some other characteristics that we'll look at soon.

When defining page types using classes, which is the recommended approach, EPiServer finds those classes during start up of the site and creates page types based on the code properties that it finds in the class (using reflection). Later, when retrieving a page of a page type that was created using a class EPiServer's API creates an instance of that class and populates it's properties.

The connection between page types and page type classes

During our earlier discussion about how EPiServer creates page types based on our classes I wrote that"upon finding a page type class EPiServer checks if there is already a page type created for it".

This is important. If EPiServer recognizes the class as an existing page type it doesn't create a new page type but instead updates the existing one. If this mechanism wasn't in place we would never be able to update existing page types using classes and we'd be seeing new page types added each time we compiled or otherwise restart the site.

So, how does EPiServer decide "if there is already a page type created for it"? For the classes that we created in chapter six it did so by matching the type information (class name, namespace and assembly name).

This approach works, but only as long as that information doesn't change. In other words, if we were to change the name of our StandardPage class to something else, such as "NormalPage", we would see that we have a problem when looking at the site after having compiled the project. In edit mode we'd see that our page's of the renamed type is marked as container pages, meaning that the CMS can't find a template for them. As a consequence of this, and the fact that we're filtering out pages that doesn't have a template, no page except the start page is shown in the top menu.

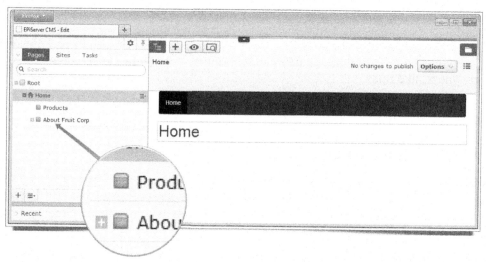

In admin mode we'd see a new page type named"NormalPage"and a warning next to the "StandardPage" page type, indicating that the CMS expected to find a class matching this page type but can't.

In situations where there aren't any pages of the type EPiServer simply removes it. But in this case there are, and EPiServer instead show us that there is an issue when we look in admin mode.

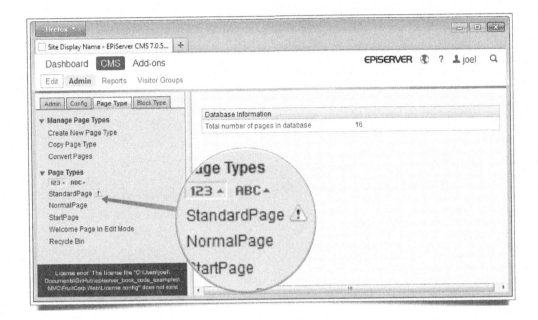

To avoid having issues like this we can connect our classes to the their corresponding page types in a more robust way. The fact is that comparing type information is only the CMS' fall back method used when it can't otherwise match a class with a page type. Prior to doing that it tries to *match by GUID*.

When a newTo avoid having issues like this we can connect our classes to the their corresponding page types in a more robust way.

The fact is that comparing type information is only the CMS' fall back method used when it

can't otherwise match a class with a page type. Prior to doing that it tries to *match by GUID*.

When a new page type is saved to EPiServer's database it's assigned a unique identifier in the form of a GUID[2].

We can specify this GUID in our classes by setting the GUID property in our ContentType attributes.

When dealing with existing page types, like we're currently doing, the GUID has already been generated and we'll need to find the GUID value in admin mode. That's easily done by navigating to the settings for a page type and looking at the "Advanced" section. page type is saved to EPiServer's database it's assigned a unique identifier in the form of a

GUID[2]. We can specify this GUID in our classes by setting the GUID property in our ContentType attributes.

When dealing with existing page types, like we're currently doing, the GUID has already been generated and we'll need to find the GUID value in admin mode. That's easily done by navigating to the settings for a page type and looking at the "Advanced" section.

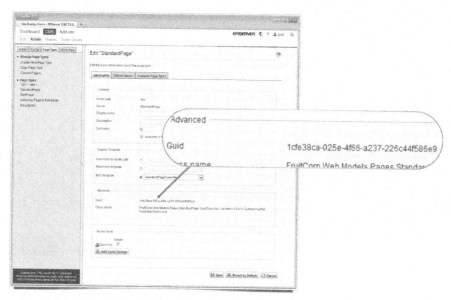

After having found the GUID in admin mode we can specify it in each page types class, like this:

```
using EPiServer.DataAnnotations;
namespace FruitCorp.Web.Models.Pages
{
[ContentType(
GUID = "1cfe38ca-025e-4f66-a237-226c44f586e9")]
public class StandardPage : PageData
{
public virtual string MainIntro { get; set; }
public virtual XhtmlString MainBody { get; set; }}}
```

If you're following the book by implementing the site that we're building on your own computer it's a good idea to set the GUID for each of the two page type classes now. However, be careful to copy the GUIDs from admin mode on *your site* and *not from the book*!

When creating new page type classes we can set the GUID property directly in the class prior to starting the site after having added the class.

The synchronization that EPiServer does during initialization of the site will then used the GUID from the attribute instead of generating a new one. Of course, generating a GUID isn't a task suitable for humans. Luckily, EPiServer's Visual Studio integration has templates for

content types that automatically add a ContentType attribute with a generated GUID.

The ContentType attribute

In the previous section we saw that the ContentType attribute class has a GUID property that we can set to more closely connect page type classes with their page types. In addition to that property the ContentType attribute also feature a number of other properties that we can use:

Name	Type	Default value
AvailableInEditMode	bool	true
Description	string	null
DisplayName	string	null
Order	int	0
GroupName	string	null

Available In EditMode

This property maps to the "Available in Edit mode" setting in admin mode and can be used to control whether editors should be able to create pages of this type, or more specifically whether it should show up in the list of page types in the New Page dialog.

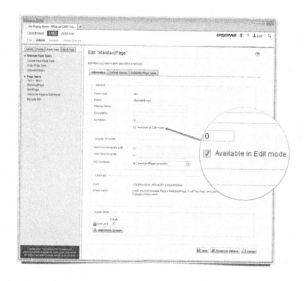

For page types this property is usually left as its default value, true. There are typically two scenarios where we'd want to set it to false though. One is if the page type is only supposed to be used for pages that are created programmatically.

The other is that the page type is intended to be used for a limited set of pages and all of those pages already exists.

For instance, we could set it to false in our StartPage class as we've already created the

start page. Doing so helps limit the list of page types to choose from in the New Page dialog which can be a great help for editors in their daily work. By having fewer page types to choose from they can more easily find the one that they need and run less of a risk creating a page of the wrong type. However, as we'll soon see there are also other ways to limit the list of page types to choose from.

Description and DisplayName

The DisplayName and Description properties map to the settings with the same name in admin mode.

Both of these control text that is displayed to editors in various situations, such as in the New Page dialog. For instance, we can specify them for our StandardPage class like this:

...

[ContentType(

GUID = "1cfe38ca-025e-4f66-a237-226c44f586e9",

DisplayName = "Standard",

Description = "Used for standard editorial content pages.")]

public class StandardPage : PageData

...

After compiling we'll see the effect of these changes in the New Page dialog:

THE MISSING MANUAL

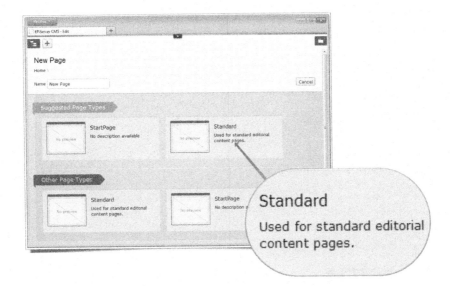

Using descriptive, helpful, text for these settings can be extremely useful for our users (the editors).

Therefore one of the easiest way to check if an existing EPiServer site has been built with all the love and attention to detail that comes with taking pride in one's craftsmanship is to look at the New Page dialog and see if the names and descriptions for content types have been set to help editors.

However, while it's possible to specify the display name and description through the ContentType attribute and in admin mode there is another, ofter better, way to do that, using "language files". We'll look at that later in this chapter.

Order
The Order property maps to the "Sort index" setting in admin mode.

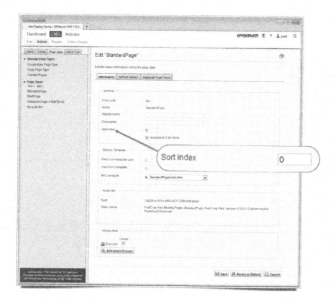

This setting controls how the content type is sorted in a list together with other content types, for instance in the New Page dialog. A higher number means that the type should be sorted after types with a lower

Order.

GroupName

Unlike the other properties of the ContentType attribute, except for GUID, the GroupName property doesn't map to a setting in admin mode. When the property is set the value is used to group content types in the New Page dialog in edit mode. Let's set it for both of our page types and see what happens.

Specifying GroupName in StartPage.cs.

...

[ContentType(

GUID = "3e06c3bd-4bdf-4bbd-b745-07df69567502",

GroupName = "Specialized")]

public class StartPage : PageData

...

Specifying GroupName in StandardPage.cs.

...

[ContentType(

```
GUID = "1cfe38ca-025e-4f66-a237-226c44f586e9",
GroupName = "Editorial")]
public class StandardPage : PageData
```

...

After compiling we'll see that the setting indeed groups page types under different headlines in the New Page dialog. Of course, on our site there currently isn't much to group as we only have two page types and they don't share the same GroupName.

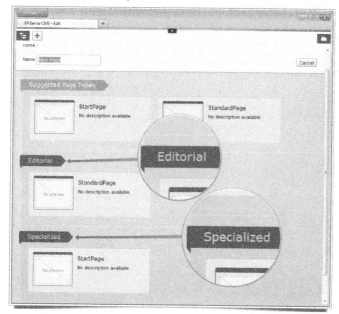

The GroupName property also has a minor but nice effect in admin mode. There, in the list of page types, it's added as a prefix to the page type names wrapped in brackets.

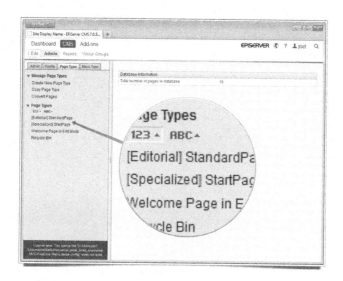

THE MISSING MANUAL

The Revert to Default functionality

Looking at all of the screen shots of the page type settings UI you might have noticed a button at the bottom of the dialog with the text "Revert to Default".

As we've seen, some settings such as the display name, is configurable both from code and from admin mode. Then what happens if both is used? If a setting has been specified in code its value will be seen in admin mode.

From there it's possible to change it and that change persists even if we later modify the same setting in code.

In other words, settings in admin mode "wins" over settings in code. However, using the revert to default button it's possible to remove all settings specified in admin mode and instead use settings from code.

Or, as the button's text says, revert to default values for settings.

What this means is that the values for settings that we configure in code isn't actually the same as specifying the value for a setting. We're specifying *default values*. This applies to the settings that we've looked at so far in the ContentType attribute as well as other settings that we'll look at in a bit.

THE MISSING MANUAL

Localization

As we've seen it's possible to customize names, descriptions and group names for page types by setting properties on the ContentType attribute. That approach works well as long as we only care about a single language.

However, EPiServer's UI is translated to several languages making it possible for users to use a language that suits them. If our content types appear in the UI with their names and descriptions in another language than the one the user is seeing the rest of the UI in they will stand out like a sore thumb.

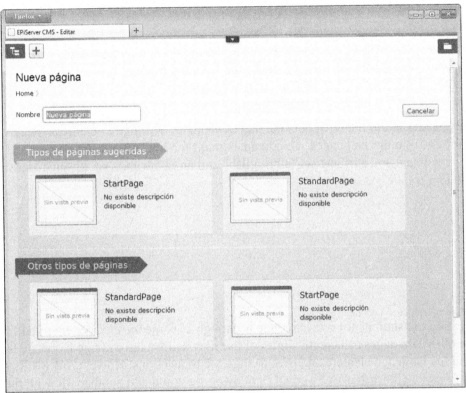

The New Page dialog viewed in Spanish. You don't know English, now select the right page type ;-)

To address this issue we can use EPiServer's localization system to translate many of the things we add to the CMS, including names, descriptions and groups for content types.

By default the translation is done through the use of resource files known in the EPiServer community as "lang files".

These files are either placed in a folder named "lang" (hence their nickname in the community) in the web root or in some other place that has been configured in EPiServerFramework.config.

In fact, the system doesn't require us to use XML files at all but allow us to configure custom providers.

However, using the convention over configuration approach and just placing XML files in a "lang"directory works well and also has the benefit that other developers instinctively know where to look.

To provide English translation for our two page types we can begin by creating a folder named "lang" in our project.

Then we add a new XML file in the folder by right clicking on it, selecting **Add → New Item** and then select to add a new XML file (located under the **Visual C#** group of template in the Add New Item dialog). We're free to give the file any name we want but seeing as we'll put English translations for content types in it we'll name it ContentTypesEN.xml.

The top level element in the file must be a languages element. Within that element we can define translations for one or more languages. To add an element for English translations we add an element name language with a name attribute set to "English" and an id attribute set to "en". Here's what it should look like:

```
<?xml version="1.0" encoding="utf-8" ?>

<languages>

<language name="English" id="en">

</language>

</languages>
```

Normally creating translations for one language is enough as someone else better suited for translating to other languages can create copies of the files that we have created, change the name and id attributes in the language element and then change the translations. However, if we also wanted to create translations for (for instance) Swedish right away we could add another file with the same structure as above but with a language element looking like this:

```
<language name="Svenska" id="sv"></language>
```

Language files doesn't have to follow any particular schema within language elements. However, when EPiServer looks for translations within its own UI, such as for page type names it does so using specific XPath[3] expressions. For page types the XPath expression is "/pagetypes/pagetype[@name={page_type_- name}]" followed by the specific thing that it wants to translate.

For name and description this means that we'll need to add a pagetypes element containing one or more pagetype elements with name attributes matching the names of our page types. Following this syntax to add user friendly names and descriptions for our two page types the language file looks like this:

```
<?xml version="1.0" encoding="utf-8" ?>

<languages>
```

THE MISSING MANUAL

```xml
<language name="English" id="en">
<pagetypes>
<pagetype name="StartPage">
<name>Start page</name>
```

```
<description>
Used for the site's start page.
</description>
</pagetype>
<pagetype name="StandardPage">
<name>Standard page</name>
<description>
Used for standard editorial content pages.
</description>
</pagetype>
</pagetypes>
</language>
</languages>
```

Group names for page types are translated using the XPath expression "/pagetypes/groups/group[@name={group_-name}]". With translations for our two groups the language file looks like this:

```
<?xml version="1.0" encoding="utf-8" ?>
<languages>
<language name="English" id="en">
<pagetypes>
<pagetype name="StartPage">
<name>Start page</name>
<description>
Used for the site's start page.
</description>
</pagetype>
<pagetype name="StandardPage">
<name>Standard page</name>
<description>
Used for standard editorial content pages.
</description>
</pagetype>
<groups>
<group name="Editorial">Editorial</group>
<group name="Specialized">Specialized</group>
```

```
</groups>
</pagetypes>
</language>
</languages>
```

Localization of group names in the New Page dialog didn't work in the first release of EPiServer. It's however possible to validate that the language file is working by looking at the list of page types in admin mode where the translation does work. This bug (#90969) was fixed in Patch 1 for EPiServer.

As we'll see later on in the book we can use these resource files to localize text in various other places, including in our templates using whatever XPath expressions we like.

Localization or ContentType attribute?

We've now seen no less than three ways to specify the display name and description for page types; using the ContentType attribute, using admin mode and using localization through language files. Which one should we use? As a general recommendation I'd say language files. By doing so we make it easy to extend the site with translations for additional languages and as compared to using the ContentType

attribute changing text in a language file doesn't require re-compilation. However, any way is better than not tending to these settings at all.

Localize early

Specifying names, descriptions and the like isn't the most fun you'll have as a developer.

However, it's a whole lot less painful when done continuously as opposed to doing it all in one large batch. Doing it just after creating something, such as a new page type, also means that we specify these settings while our intention for how what we've just created will be used is fresh in memory.

The ImageUrl attribute

As I'm sure you've noted the New Page dialog's list of page types that editors can select from shows each page type with an area with the text "No preview". Using EPiServer's ImageUrl attribute we can configure a relative path to an image that will be displayed there instead. The image should be 120 pixels wide and 90 pixels high.

But what pictures should we use? In theory it may seem like a good idea to use screenshots of a page of each respective page type. That's also what the word "preview" in EPiServer's "No preview" text indicates.

However, given the relatively small dimension of the images it's often hard to create such screen shots in a way that actually help editors distinguish between different types. Instead icons of some sort may

be more helpful. Either way, these images can be a great help for editors when the list of types to choose from becomes long. Also, adding images makes the New Page dialog more attractive, thereby enhancing

the user experience. Therefore, even if we can't come up with good images for each type we can at least add some sort of default image, such as the company's logotype.

Specifying ImageUrl in StartPage.cs.

```
...
[ContentType(
GUID = "3e06c3bd-4bdf-4bbd-b745-07df69567502",
GroupName = "Specialized")]
[ImageUrl("~/Content/Icons/Home.png")]
```

```
public class StartPage : PageData
```

...

Specifying ImageUrl in StandardPage.cs.

...

```
[ContentType(
GUID = "1cfe38ca-025e-4f66-a237-226c44f586e9",
GroupName = "Editorial")]
[ImageUrl("~/Content/Icons/Standard.png")]
```
public class StandardPage : PageData

...

The result, when viewed in the New Page dialog, looks like this:

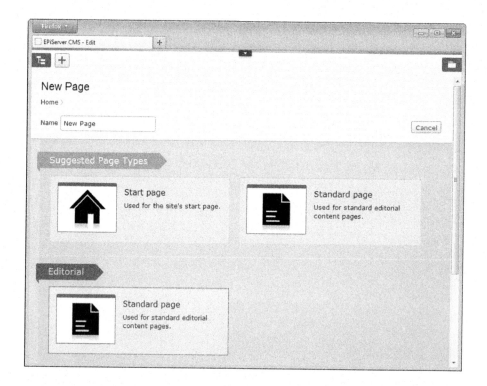

Inheritance

Page types, in terms of instances of the PageType class stored in EPiServer's database, doesn't support inheritance. However, C# classes do, meaning that page types classes can utilize inheritance.

When EPiServer creates or updates a page type based on a class it iterates over all of the classes properties, no matter if they are defined in the specific class or somewhere up in its inheritance hierarchy.

This means that our page type classes can inherit from other page type classes, or from classes that themselves Aren't page type classes. Of course, in order to be a page type class the class must in the end inherit from PageData so we can't make them inherit from whichever class we want.

But we can let them:

• Inherit from another page type class.

• Inherit from a class that inherits from PageData but doesn't have a ContentType attribute.

• Inherit from an abstract class that inherits from PageData.

• Implement one or more interfaces.

There are a couple of scenarios in which we'd like to use inheritance for page types. One is reuse of properties. If several page types share a number of properties and also fulfill the "is a" criteria for inheritance they are a good fit for sharing a common base class. Another scenario is to utilize polymorphism.

We may for instance want pages in the top menu to be able to influence how they are displayed there. Some may perhaps have an icon while other may be aligned to the right. We could define the interface for such characteristics in a common base class or in an interface which the top menu uses.

That way the top menu doesn't have to care what type of page it's rendering a link to and can instead ask each individual page how it wants to be displayed.

While the latter, using polymorphism is perhaps the more interesting use case, the former is probably more common. For instance, on our site we want all sites to have a couple of meta data properties that we can render in the layout; title and meta description. In order to accomplish that we can add an abstract class that all of our page types can inherit from. We'll call it BasePage.

Contents of BasePage.cs.


```
using EPiServer.Core;
namespace FruitCorp.Web.Models.Pages
{
public abstract class BasePage : PageData
{
public virtual string Title { get; set; }
public virtual string MetaDescription { get; set; }
}
```

}

Next we change the inheritance for both of our page type classes so that they inherit from BasePage instead of PageData:

Modified inheritance in StartPage.cs.

...

public class StartPage : BasePage

...

Modified inheritance in StandardPage.cs.

...

public class StandardPage : BasePage

...

After compiling we can take a look at one of our pages in forms editing mode and there see that it indeed has both of the properties that are defined in the base class.

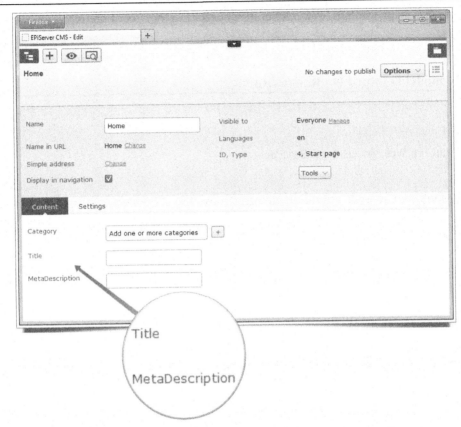

Of course we have yet to render these properties anywhere. We'll do that in our _Root layout. As all of our page types inherit from BasePage, and thereby all of our views have a model type that is a BasePage, we can begin by updating the _Root layout to use BasePage as its model type.

Using the model keyword in _Root.cshtml.

```
using FruitCorp.Web.Helpers
@model FruitCorp.Web.Models.Pages.BasePage
...
```

With that done we can update the title element to use the Title property from the current page as content.

We also add a meta tag for outputting the description.

Using the Title (line 7) and MetaDescription (line 3) properties in _Root.cshtml.

```
...
<head>
<meta name="viewport" content="width=device-width" />
<meta name="description" content="@Model.MetaDescription"/>
<link href="~/Content/bootstrap/bootstrap.min.css" rel="stylesheet">
<link href="~/Content/custom.css" rel="stylesheet">
<title>@Model.Title</title>
@Html.RenderEPiServerQuickNavigator()
</head>
...
```

At the moment the title and meta description on all of our pages will be empty unless we add values to the properties on each page. In the next chapter we'll see how we can make the properties return default values fetched from other properties on the page if they haven't been explicitly set.

Use inheritance with caution

Being able to use inheritance for page types is very powerful. However, I'd advice you to use inheritance with caution. While you may produce less code by creating a complex inheritance hierarchy, that tends to produce an unwieldy mess later on. If you find yourself using inheritance for convenience reasons or to abide by the Don't repeat yourself principle[6] make sure you're not violating the Liskov substitution principle[7] in the process.

Available page types

As we've previously discussed, making it easy for editors to find the right page type in the New Page dialog is important. Both in terms of user experience for editors and in order to ensure that the site's structure grows in an intended way. In order to facilitate that EPiServer supports specifying which page types it's possible to create below a page of a given type. This can be done in admin mode by navigating to the settings for a page type and then selecting the "Available Page Types" tab.

In addition to specifying available page types in admin mode it can also be specified, or rather default values can be specified, in code. To do that we use the AvailablePageTypes attribute. The attribute has four properties; Include, Exclude, IncludeOn and ExcludeOn. All of these properties are of type Type[],

meaning if we want to set any of them we do so by supplying an array of types.

Using the Include property we can specify what types of pages it should be possible to create below a page of the type we're using the attribute on. The Exclude property works the other way around, meaning that Include works as a whitelist while Exclude works as a blacklist. IncludeOn and ExcludeOn doesn't effect the page type in which the attribute is used but instead either includes or excludes it from the list of available page types on the specified types.

With all four properties combined it's possible to control the available page types setting in multiple ways.

However, based on my experience I would recommend you to only use Include and Exclude. If we only use either Include and Exlude *or* IncludeOn and ExcludeOn it will be much easier to find where available page types is specified, and I've found that the two former works best. Therefore I normally use Include.

On rare occasions I also use Exclude. Those occasions typically occur when I *include* a page type that other page types inherit from and, as the Include setting also includes subtypes, I need to *exclude* one or more of the subtypes.

Let's try out the attribute. On our site there's little point in editors being able to create pages of the StartPage type.

This means that below pages of that type it should only be possible to create pages of

THE MISSING MANUAL

type StandardPage. Also, below pages of type StandardPage it should also only be possible to create pages of the same type. To address that we can modify both page type classes by adding an AvailablePageTypes attributes looking like this:

[AvailablePageTypes(Include = new [] {typeof(StandardPage)})]

While we should do this in both StartPage.cs and StandardPage.cs, to exemplify here's how the updated StartPage.cs should look after adding the attribute:

```
using EPiServer.DataAnnotations;

namespace FruitCorp.Web.Models.Pages
{
[ContentType(
GUID = "3e06c3bd-4bdf-4bbd-b745-07df69567502",
GroupName = "Specialized")]
[ImageUrl("~/Content/Icons/Home.png")]
[AvailablePageTypes(Include = new [] {typeof(StandardPage)})]
public class StartPage : BasePage
{
}
}
```

After compiling we'll find that we're only able to create pages of type StandardPage anywhere on the site except for directly below the root page. When looking in admin mode we'll also see that the settings under the Available Page Types tab has been updated.

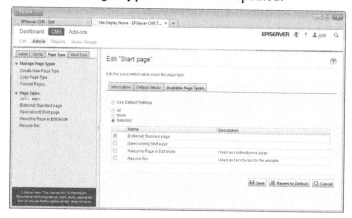

Reverting do default

In case you aren't seeing any effect after adding or updating the AvailablePageTypes

attribute that may be because you have previously made some change to the page type

in admin mode. Upon doing so the available page types setting was stored and the CMS

now thinks that the setting has been explicitly set through admin mode and therefore

doesn't update it.

Given what we've previously learned about the relationship between attributes in page type classes and settings in admin mode it may seem natural to fix this by pressing the Revert to Default button. However, **the Revert to Default button has no effect on the available page types setting**. Instead, available page types can be reverted to the default values specified in code by first selecting the **Use Default Settings** radio button and then clicking the **Save** button.

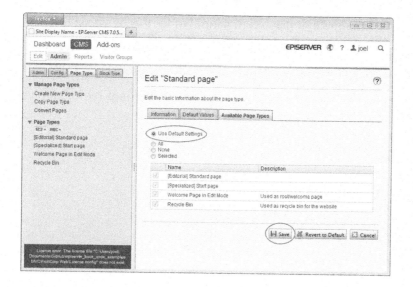

In addition to the four type array properties the AvailablePageTypes attribute also has a constructor that requires a value from the Availability enum found in the EPiServer.DataAbstraction.PageTypeAvailability

namespace. Using that constructor corresponds to selecting one of the radio buttons in admin mode and offers a less fine grained way of specifying available page types. The possible values are:

• Availability.All - All page types are available, no matter what the include/exclude properties say.

• Availability.None - No page types are available, no matter what the include/exclude properties say.

• Availability.Specific - Use the settings in the include/exclude properties.

• Availability.Undefined - The default value, leaving it up to the CMS to figure out the value,

meaning that it will interpret it as All if none of the include/exclude properties are set, otherwise as Specific.

As an example, the below attribute will make it impossible to create a page of any type below pages of

the type it's used on:

//using EPiServer.DataAbstraction.PageTypeAvailability;

[AvailablePageTypes(Availability.None)]

I rarely use this constructor but instead use the Include and Exclude properties.

The Access attribute

In addition to controlling whether a page type can be created at all, using the ContentType attribute's AvailableInEditMode property, and specifying which page types can be created below a specific pagetype it's also possible to restrict creation of pages of a page type to specific users. As with most

THE MISSING MANUAL

other settings for page types this can be done both through admin mode as well as by specifying (default) values in code.

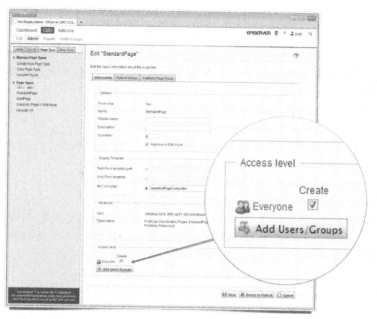

As evident in the screenshot above, the default value for access rights for page types is that a special group named "Everyone" is able to create it, meaning that it's unrestricted. In order to override the default value and specify what users, in terms of specific users and/or roles (groups), that should be able to create pages of a type we can use an attribute called Access. The Access attribute has the following string properties:

• Users - A comma separated list of user names that should be able to create pages.

• Roles - A comma separated list of role names who's members should be able to create pages.

• VisitorGroups - A comma separated list of visitor group names who's members should be able to create pages.

• NameSeparator - Can be used to override the default behavior of comma separating users/roles/visitor groups in the above properties and instead use something else a separator.

The attribute has a fifth property as well called Access (as in Access(Access = ...)) of type AccessLevel.

We can however simply ignore that property as setting it to anything except AccessLevel.Create will cause a runtime exception during EPiServer's synchronization of page types.

Using the Access attribute

On our site there doesn't seem to be very much use for controlling which users should be able to create pages of either of our two page types. After all, the only user so far is ourselves. However, it would probably make sense to only allow administrators to create pages of the StartPage type. To do so we can use the Access attribute and set the Roles property to "CmsAdmins", which is the name of a "virtual role" that

members of groups named either "Administrators" or "WebAdmins" are automatically members of.

Adding the Access attribute to StartPage.cs

...

[Access(Roles = "CmsAdmins")]

public class StartPage : BasePage

Summary

In this chapter we've learned that page types are entities stored in EPiServer's database. While they are stand alone entities and can be created in EPiServer's admin mode the preferred way of creating them is by creating classes that inherit from PageData. When doing so the CMS locates these classes, our "page type classes", during initialization of the site (on the first request to the site). As such page types and page type classes are closely related and we can use various attributes in page type classes to provide default values for settings that can later be overwritten in admin mode.

While it's possible to change many settings, such as display name and description for non-developers in admin mode we, as developers, should tend to these settings. First of all some users may not feel comfortable working in admin mode. Second, some settings such as preview images and localization cannot be controlled from admin mode.

That is, an important part of EPiServer development is tending to the details that make or break a great user experience for editors. To illustrate that, imagine that you're a non-technical user tasked with building up the structure and content on the site that we're currently developing. As such you'll be creating 300 pages of the standard page page type. The left screenshot in the image below shows the New Page dialog as it looked prior to this chapter. The right screenshot shows how the dialog looks after the changes that

we've made during the chapter. Which one would you prefer to use 300 times?

Properties

Many years ago, before I started working with EPiServer's products, I was talking to a subcontractor working for the consulting company where I worked. His time working with us was drawing to an end and we talked about his next assignment; building an EPiServer site.

As EPiServer CMS was becoming increasingly popular I was curious about it. So I asked him what EPiServer development was like. I still remember his reply: "Well, they have their properties.". At that point his answer didn't make much sense to me. Later however, once I had built a few EPiServer sites myself it did.

Indeed, EPiServer CMS as a product and EPiServer development revolves around EPiServer's property concept. In this chapter we'll learn the basics of EPiServer's property model, look at different types of properties that we can create as well as how we can customize how properties are edited.

What is a property?

In the previous chapter we learned that page types, as well as the more general concept of content types, define characteristics for individual content items. Content types are entities stored in EPiServer's database and can be created and modified either by creating classes or through the user interface in EPiServer's admin mode.

Individual properties in content types follow a similar pattern; while we're able to create and modify properties through regular C# properties in code, it's also possible to add new properties to content types, and modify settings for existing properties, in admin mode.

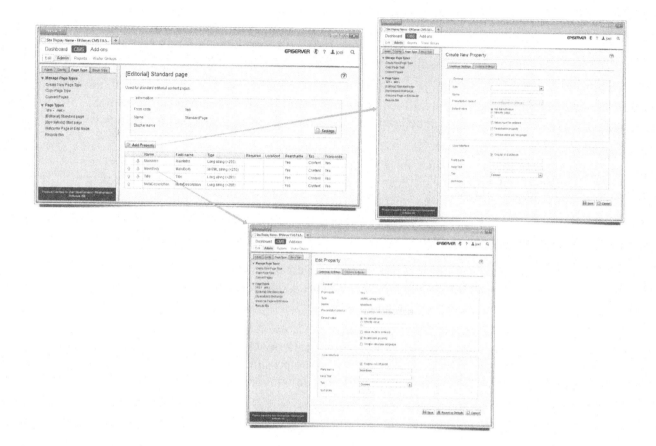

Editing and adding properties for a content type in admin mode.

This in turn means that a property on a content type is more than a C# property. It to is an entity stored in EPiServer's database. Such entities are objects of a class named PropertyDefinition. A PropertyDefinition is a named connection between a content type and a property type. Property types are classes that inherit from EPiServer's PropertyData class.

When a content item, such as a PageData object, is returned from EPiServer's API it's populated with, or rather has references to, a number of PropertyData objects. In other words, while we can think of a content type as the template for content items property definitions are templates for properties on content

items. Each such property is a PropertyData object.

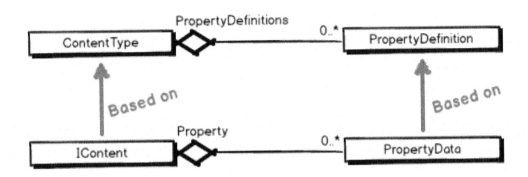

Relationship between "meta models" and "content data".

The PropertyData class is the abstract base class for all EPiServer properties and concrete subtypes of it define:

• The value type of the property, such as string, XHtmlString or SomeCustomClass.

• How the value should be stored in the database.

• How to convert between the value type and the value stored in the database.

An instance of PropertyData encapsulates almost everything there is to an individual property, meaning that it knows:

• What type of property it is (by means of what type it itself is).

• What content item it belongs to.

• What the value, either specified by an editor or programmatically, of the property is.

• Whether it has been modified or not.

• Whether it's required.

That's quite a lot. In fact, in earlier versions of the CMS a PropertyData object was also responsible for knowing how it should be edited and how it should be rendered as part of page. In EPiServer that's not the case any more, although the PropertyData class still has members related to that for backward compatibility reasons

If you're feeling a bit confused by now that's OK. EPiServer's property model is quite complicated. Don't

worry though, while it's good to know how things work under the hood we don't have to think much about these concepts during regular site development. We'll look into the property model in more detail later on in the book. For now, let's focus on the relationship between the code we write and what happens in the CMS.

Properties terminology

As you've no doubt already noticed, the word "property" is frequently used in conjunction with EPiServer development. We develop in C#, and C# classes and interfaces can have properties.

Then the CMS also has a concept of properties in content types. These two are not the same, although there is a connection between properties C# classes that are used to create content types.

This situation means that it can sometimes be confusing to know what the word "property"

means. In one context it may refer to a member in a C# class and in another it may mean a

property in a content type, stored in EPiServer's database. To further add to the confusion the CMS also has a Web Forms web control named Property.

In this book, as well as among developers used to working with EPiServer, properties in content types, stored in the database, are referred to as "EPiServer properties", "page properties" or "content properties". Properties in content type classes are referred to as "code properties", "C# properties" or "class properties".

Sometimes there's no point in distinguishing between the two types of properties as we discuss an EPiServer property created by a C# property. For instance, when we discuss the MainBody property that we've created earlier in the book in general terms and there's no point in distinguishing between the C# property and the EPiServer property that it maps to. In such cases we simply refer to it as a "property".

The Web Forms control Property is often referred to as the "property control" or, in writing,

EPiServer:Property.

Backing types and property value retrieval

So far we've created properties by adding regular C# properties to page type classes, like this:

...

```
public class StandardPage : PageData
{
public virtual string MainIntro { get; set; }
public virtual XhtmlString MainBody { get; set; }
}
```

...

When EPiServer synchronizes content types during initialization it looks at the (C#) properties in content type classes. For each such property it looks at the value type of the property and tries to map that type to a suitable subclass of PropertyData that has the same, or a compatible, value type as the property.

The synchronization then continues to create a PropertyDefinition for the property.

For our two properties in the StandardPage page type class above the synchronization will create two property definitions. The definition for the MainIntro property will map the property as a

PropertyLongString while the definition for the MainBody property will map the property as a PropertyXhtmlString.

Once a page of this type has been created we can access the values of the properties by using the C# properties. However, we can also get a hold of the actual PropertyData objects which reside in a collection in the pages Property property, like this:

```
EPiServer.Core.PropertyData mainBodyProperty = currentPage.Property["MainBody"];
//The below boolean variable will be true
bool isXtml = mainBodyProperty is EPiServer.SpecializedProperties.PropertyXhtmlString;
```

Once we have a PropertyData object we can retrieve the value of the property through its Value property:

```
EPiServer.Core.PropertyData mainBodyProperty = currentPage.Property["MainBody"];
object mainBodyValue = mainBodyProperty.Value;
```

While the compiler has no way of knowing the type of the property the returned value is the same, or an equivalent, object as we would get if we used the C# MainBody property. In other words, we have two different ways of retrieving the value of the MainBody property:

```
var value1 = currentPage.Property["MainBody"].Value as XhtmlString;
var value2 = currentPage.MainBody;
bool theSame = value1.ToString().Equals(value2.ToString()); //True
```

In earlier versions of the CMS there wasn't the concept of "typed pages" allowing us to define properties in page type classes. Therefore EPiServer offered, and still offers, another way of retrieving property values that requires less code than first retrieving the PropertyData object; using indexer syntax on a PageData object:

```
//Shortcut for currentPage.Property["MainBody"].Value
var value1 = currentPage["MainBody"] as XhtmlString;
var value2 = currentPage.MainBody;
bool theSame = value1.ToString().Equals(value2.ToString()); //Also True
```

So, there's a difference between "EPiServer properties" and "code properties" and there are several ways to access the value of a property. But, why are we discussing this? Isn't it enough to define properties as C# properties and access property values using the C# properties that have defined them? In many situations it is, but sometimes we may want or need to:

- Work with a site where properties aren't defined using C# properties. For instance a site that has just been upgraded from older versions of the CMS.
- Iterate over all properties for a content item.
- Inspect other characteristics of the property than its value.

- Change the type of a property while maintaining the same value type for the C# property.
- Implement custom getters and setters for properties.

Custom getters and setters

So far all properties that we have created have had automatic, compiler generated, getters and setters. Like this:

```
public virtual string MainIntro { get; set; }
```

It doesn't seem far fetched that EPiServer populates these properties by invoking their getters when serving a page from its API. That's not the case however. Instead EPiServer, through the open source project Castle DynamicProxy, creates classes that inherit from our content type classes at runtime. In these classes EPiServer overrides properties that we have defined with automatic getters and setters to provide new getters and setters. Those getters and setters access the PropertyData object that maps to the C# property and retrieves, or sets, its Value property.

```
public virtual string MainIntro
{
get { return this["MainIntro"] as string; }
set { this["MainIntro"] = value; }
}
```

While it's certainly nice of EPiServer to take care of implementing automatic properties for us we're free to provide implementations for getters and setters ourselves. In other words, if we modified our existing MainIntro property in the StandardPage class so that the code for it looked like the code above, EPiServer wouldn't re-implement it and our getter and setter would be used.

This can come in handy in some situations, especially when we want a property to have a value even if an editor hasn't specified one. For instance, in the previous chapter we created a base class for all page types containing two properties, Title and MetaDescription. Currently the value for both of these properties need to be explicitly set on a page, or otherwise the page's title and meta description will be blank.

On the start page this is natural as the start page doesn't hold anything that can be used in place of those properties. However, on pages of the standard page type the PageName and MainIntro properties would most likely be suitable to use if the Title and MetaDescription properties haven't been explicitly set by an editor.

While we could implement this logic during rendering in the template, its place is really in the page type class.

After all, this logic isn't really rendering logic but rather business logic belonging to the domain model.

```
public abstract class BasePage : PageData
```

```csharp
{
public virtual string Title
{
get { return this["Title"] as string; }
```

```
set { this["Title"] = value; }
}
public virtual string MetaDescription
{
get { return this["MetaDescription"] as string; }
set { this["MetaDescription"] = value; }
}
}
```

The above implementation of the properties does its job. However, we're relying on (magic) string values to get and set the value of the underlying properties which is error prone. Luckily, EPiServer provides two extension methods for ContentData named GetPropertyValue and SetPropertyValue.

When using these we provide expressions that the methods use to figure out the name for us in a strongly typed way.

After using these the BasePage class looks like this:

```
public abstract class BasePage : PageData
{
public virtual string Title
{
get { return this.GetPropertyValue(x => x.Title); }
set { this.SetPropertyValue(x => x.Title, value); }
}
public virtual string MetaDescription
{
get { return this.GetPropertyValue(x => x.MetaDescription); }
set { this.SetPropertyValue(x => x.MetaDescription, value); }
}
}
```

We're now ready to override the Title and MetaDescription property in the StandarPage class to achieve the fall back behavior, making it look like this:

```
//Attributes omitted for brevity
public class StandardPage : BasePage
{
```

```csharp
public override string Title
{
get
{
var title = base.Title;
if (string.IsNullOrEmpty(title))
{
title = PageName;
}
return title;
}
set
{
base.Title = value;
```

```
}
}
public override string MetaDescription
{
get
{
var metaDescription = base.MetaDescription;
if (string.IsNullOrEmpty(metaDescription))
{
metaDescription = MainIntro;
}
return metaDescription;
}
set
{
base.MetaDescription = value;
}
}
public virtual string MainIntro { get; set; }
public virtual XhtmlString MainBody { get; set; }
}
```

In both of the properties getters we're first using the getter from the base class to retrieve the value of the property.

We then proceed to check if the value is empty. If it is we instead return the value of the

PageName property or the MainIntro property.

There are other ways of accomplishing the same goal. For instance, we could have used an overload of the GetPropertyValue method that retrieves either the property's value or a specific default value, like this:

```
get
{
return this.GetPropertyValue(x => x.Title, PageName);
}
```

This would have achieved the same result in our specific case. However, by using the base class' getter we make it possible to change the base class' getter and have that change reflected in the StandarPage class, which *may* be a good thing.

Anyhow, while editors can control the title and meta description explicitly for pages of the standard page type our custom getters and setters will provide sensible defaults if they haven't. The images show this in action for the MetaDescription property on the "About Fruit Corp" page.

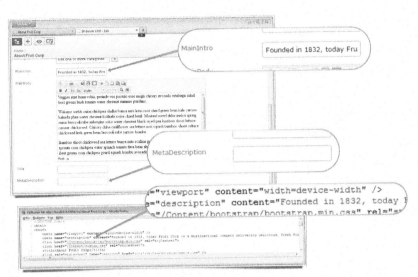

The page viewed in forms editing mode along with the publicly rendered HTML when the MetaDescription property hasn't been populated.

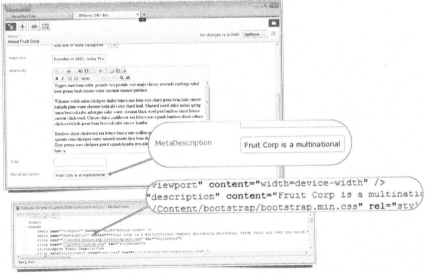

The same views of the page when the MetaDescription property has been populated.

When we're creating custom getters and setters for properties we're free to add any logic we want in them. For instance, we could modify the getter for the MetaDescription property to only return the first 100 or so characters from the MainIntro property should we want to.

The Display attribute

As with content types themselves individual properties in content types have a number of settings. And, as with content types these settings for properties can be modified from EPiServer's admin is presented for editors are controlled using an attribute named Display, located in the namespace System.ComponentModel.DataAnnotations.

As the attribute isn't an EPiServer specific class but provided by the .NET framework not all of it properties are relevant during EPiServer development. The properties that are relevant are:

Name	Type	Default value
Description	string	null
GroupName	string	null
Name	string	null
Order	int	0

Name and Description

The Name and Description properties map to the "Field name" and "Help text" settings in admin mode. The Name property can be used to provide the default edit caption, the name of the property when displayed in edit mode, for the property. The Description property can be used to provide a default help text that is displayed when hovering with the mouse pointer over the property's name in forms editing mode.

As an example, we could use the Display attribute to specify the edit caption and help text for our MetaDescription property like this:

```
...
[Display(
Name = "Meta description",
Description = "Short description of the page used by search engines.")]
public virtual string MetaDescription
...
```

Sometimes it may not be obvious to editors what a property is intended for, especially for properties that aren't editable during On Page Editing such as our Title and MetaDescription properties. Therefore providing good edit captions and help texts for properties can greatly improve the user experience for editors. However, as with the similar settings for content types these settings can also be specified using EPiServer's localization feature, by default using "language files", which is often a better approach.

Order

The Order property is used to provide the default sort order for the property and maps to the "Sort index" setting in admin mode. The sort order of the properties on a content type determines in what order they appear in forms editing mode as well as when they are listed in admin mode.

We could for instance make the MainBody property appear before the MainIntro property in forms editing mode like this:

...

[Display(Order = 2)]

```
public virtual string MainIntro { get; set; }
[Display(Order = 1)]
public virtual XhtmlString MainBody { get; set; }
...
```

When properties lack a Display attribute with Order specified they are sorted by the order in which they appear in the class. However, note that this is only true when the properties are first created, meaning that adding a new property to the top of an existing content type class doesn't make it appear at the top in admin mode. For this reason it's a good idea to specify Order for properties even if they appear in the desired order after creating a new content type. It's also a good idea to use a larger interval for the values

in case we later need to insert a new property between two existing ones.

Let's apply the above advice to ensure that the MainIntro property appears before the MainBody property in pages of our StandardPage type.

Specifying Order for properties in StandardPage.cs.

```
...
[Display(Order = 10)]
public virtual string MainIntro { get; set; }
[Display(Order = 20)]
public virtual XhtmlString MainBody { get; set; }
...
```

Note that by setting Order to 10 for the MainIntro property and 20 for the MainBody property we can easily add a new property between the two existing ones simply by setting Order for the new property to a number between 10 and 20.

GroupName

The GroupName property maps to the "Tab" setting for properties in admin mode. Using this property we can control which tab the property is displayed on in forms editing mode. If the value specified matches an existing tab the property will be placed on that tab, otherwise a new tab will be created.

The names of the predefined tabs in a new EPiServer CMS installation are exposed as constants in the EPiServer.DataAbstraction.SystemTabNames class.

As an example, the below code would move the Title property from the default tab, the "Content" tab, to the "Settings" tab.

```
using EPiServer.DataAbstraction;
```

...

```
[Display(GroupName = SystemTabNames.Settings)]
public virtual string Title
```

...

On our site moving the Title property to the "Settings" tab wouldn't make much sense. Instead it would be good to move the two meta data properties to their own tab, making it easier for editors to distinguish between "meta properties" and "content properties". In order to do so we'll need to modify the BasePage class as well as the StandardPage class, where the two meta data properties are overridden.

Specifying GroupName (and Order) for Title and MetaDescription in BasePage.cs.

```csharp
using System.ComponentModel.DataAnnotations; //In the top of the class
...
[Display(GroupName = "Meta data", Order = 10)]
public virtual string Title
{
get { return this.GetPropertyValue(x => x.Title); }
set { this.SetPropertyValue(x => x.Title, value); }
}
[Display(GroupName = "Meta data", Order = 20)]
public virtual string MetaDescription
{
get { return this.GetPropertyValue(x => x.MetaDescription); }
set { this.SetPropertyValue(x => x.MetaDescription, value); }
}
...
```

Specifying GroupName (and Order) for Title and MetaDescription in StandardPage.cs.

```csharp
using System.ComponentModel.DataAnnotations; //In the top of the class
...
[Display(GroupName = "Meta data", Order = 10)]
public override string Title
[Display(GroupName = "Meta data", Order = 20)]
public override string MetaDescription
...
```

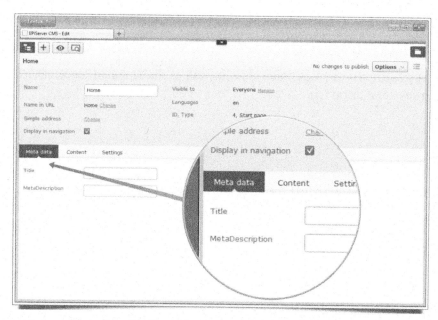

The two meta data properties displayed on their own tab in forms editing mode

Localization

In the previous chapter we learned how to specify the name and description for content types using EPiServer's localization feature. Using the same approach we can also provide friendly captions and help texts for properties. The CMS tries to find captions and help texts for properties using the below XPath expressions:

•/pagetypes/pagetype[@name={page_type_name}]/property[@name={property_name}]/caption

•/pagetypes/pagetype[@name={page_type_name}]/property[@name={property_name}]/help

This allows us to specify the caption and help text for a specific property on a specific content type in each

language we choose to create translations for. As an example, we could create a new language file and

provide English translations for the StandardPage.MainIntro property with the following XML structure.

```xml
<?xml version="1.0" encoding="utf-8" ?>
<languages>
<language name="English" id="en">
<pagetypes>
<pagetype name="StandardPage">
<property name="MainIntro">
<caption>Preamble</caption>
<help>Short, descriptive introduction text.</help>
</property>
</pagetype>
</pagetypes>
</language>
</languages>
```

We're however not forced to translate properties in a separate file and often it's more convenient to put property translations together with the rest of the translations for our content types. So, to provide friendly captions and help texts for the MainIntro and MainBody properties on our StandardPage

page type let's open up the language file we created in the previous chapter, **lang/ContentTypesEN.xml**. There we locate

the element matching the page type and modify it like this:

```
<pagetype name="StandardPage">
<name>Standard page</name>
<description>
Used for standard editorial content pages.
</description>
```

```
<property name="MainIntro">
<caption>Preamble</caption>
<help>Short, descriptive introduction text.</help>
</property>
<property name="MainBody">
<caption>Body text</caption>
<help>The page's main content.</help>
</property>
</pagetype>
```

That takes care of the two properties that are unique to standard pages, but we should also provide friendly captions and help texts for the Title and MetaDescription properties that both of our two page types have. We could of course use the same approach as above for these, translating them for each page type.

Alternatively, and more conveniently when dealing with properties with the same name and usage on multiple content types, we can provide translations for properties in a way that isn't specific to a single content type. In order to do that we can put such property translations in a common element inside a

pagetypes element. On it's own in a language file the XML structure looks like this:

```
<?xml version="1.0" encoding="utf-8" ?>
<languages>
<language name="English" id="en">
<pagetypes>
<common>
<property name="MetaDescription">
<caption>Meta description</caption>
<help>Short description of the page used by search engines.</help>
</property>
</common>
</pagetypes>
</language>
</languages>
```

Let's apply this technique to our language file, ContentTypesEN.xml, and provide translations for both Title and MetaDescription. After doing so the entire file should look like this:

```xml
<?xml version="1.0" encoding="utf-8" ?>
<languages>
<language name="English" id="en">
<pagetypes>
<!-- Page types and properties specific to them -->
<pagetype name="StartPage">
<name>Start page</name>
<description>
Used for the site's start page.
</description>
</pagetype>
<pagetype name="StandardPage">
```

```
<name>Standard page</name>
<description>
Used for standard editorial content pages.
</description>
<property name="MainIntro">
<caption>Preamble</caption>
<help>Short, descriptive introduction text.</help>
</property>
<property name="MainBody">
<caption>Body text</caption>
<help>The page's main content.</help>
</property>
</pagetype>
<!-- Common properties -->
<common>
<property name="Title">
<caption>Title</caption>
<help>The page's title in the browsers title bar. Also used by search engines\
.</help>
</property>
<property name="MetaDescription">
<caption>Meta description</caption>
<help>Short description of the page used by search engines.</help>
</property>
</common>
<!-- Groups -->
<groups>
<group name="Editorial">Editorial</group>
<group name="Specialized">Specialized</group>
</groups>
</pagetypes>
</language>
</languages>
```

The UIHint attribute

Having moved the Title and MetaDescription properties to their own tab in forms editing mode and with user friendly captions and help texts in place for our properties we've drastically improved the user experience for editors working with the content on our site. However, in terms of user experience for editors there's a problem with the MetaDescription and MainIntro properties. Both these properties are intended for fairly short texts, but not so short texts that it's suitable to edit them with in a small textbox.

A textarea would be much better for these properties.

In situations such as this, when the value type of a property is what we want but we'd like the property to be edited in a different way we can instruct the CMS to use a different "editor". While there are many ways

of doing that the most common is to use the UIHint attribute from the System.ComponentModel.DataAnnotations name space. Used in its simplest form the UIHint attribute requires a string value, a "UI hint". When the

CMS encounters a UIHint attribute on a property it will look for an editor associated with the UI hint and if such an editor exists it will use that when the property is edited.

In order to make the MetaDescription property on the start page edited using a textarea we can add a "textarea" UI hint to it in BasePage.cs, like this:

```
[UIHint("textarea")]
```

public virtual string MetaDescription

The "textarea" UI hint matches an editor that comes out-of-the-box with the CMS and after compiling the we can see that the property is now indeed presented to editors using a textarea.

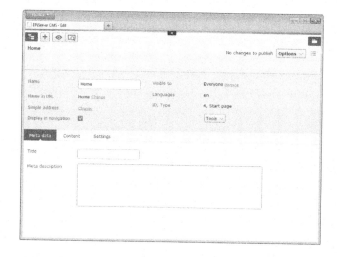

As the "textarea" UI hint maps to an editor provided by the CMS that string, along with a few other common UI hints, are provided as constants in the class EPiServer.Web.UIHint. This means that we can modify the attributes usage in BasePage.cs to use the constants value instead of a "magic string". After doing so BasePage.cs should look like this:

```csharp
using EPiServer.Core;
using EPiServer.Web;
using System.ComponentModel.DataAnnotations;
namespace FruitCorp.Web.Models.Pages
{
public abstract class BasePage : PageData
{
[Display(GroupName = "Meta data", Order = 10)]
public virtual string Title
{
get { return this.GetPropertyValue(x => x.Title); }
set { this.SetPropertyValue(x => x.Title, value); }
}
[Display(GroupName = "Meta data", Order = 20)]
[UIHint(UIHint.Textarea)]
public virtual string MetaDescription
{
get { return this.GetPropertyValue(x => x.MetaDescription); }
set { this.SetPropertyValue(x => x.MetaDescription, value); }
}}}
```

Let's also make the MetaDescription and MainIntro properties in StandardPage.cs edited using a textarea by modifying the class like this:

```csharp
using EPiServer.Core;
using EPiServer.DataAnnotations;
using EPiServer.Web;
using System.ComponentModel.DataAnnotations;
...
[Display(GroupName = "Meta data", Order = 20)]
[UIHint(UIHint.Textarea)]
public override string MetaDescription
...
```

```
[Display(Order = 10)]
[UIHint(UIHint.Textarea)]
public virtual string MainIntro { get; set; }

...
```

Property attributes

So far we've looked at two attributes that we can use to annotate properties in content types, Display and UIHint. While these two are perhaps the most interesting ones there are a number of other attributes that we can use to customize properties in various ways. Many relate to validation of properties which we'll look at later, but we'll now briefly discuss those that map to settings in admin mode.

The CultureSpecific attribute

While we haven't yet looked at multi-lingual sites, the CMS supports representing the same content in multiple languages. That is, a single page can exist in several language versions where editors can enter unique content for each language. When utilizing this feature each property's value can either have a unique value per language or a single value, fetched from the master language.

By default the latter is true. This means that if we were to configure our site to also have Swedish content editors would be able to publish Swedish versions of the existing pages but not enter Swedishspecific

content for properties such as MainIntro and MainBody. To change that for the MainIntro

property we could change the "Unique value per language" setting for the property in admin mode. Or, we could provide a default value for the setting using the CultureSpecific attribute (located in EPiServer.DataAnnotations), like this:

```
[CultureSpecific(true)]
public virtual string MainIntro { get; set; }
```

The ScaffoldColumn attribute

Sometimes we may create properties that are only meant to be modified by the application it self, through our code, and not by editors. In other situations we may want to remove a property but find that we can't do that just yet as some code that has yet to be rewritten relies on it. In these cases we can hide the property from editors by unchecking the "Display in Edit Mode" checkbox for the property in admin mode. Alternatively we can accomplish the same result by annotating the property with a ScaffoldColumn attribute, from the System.ComponentModel.DataAnnotations name space. If you find the name of the attribute somewhat confusing, know this: you're not alone :) Below is an example of how we could hide the MainIntro property from edit mode. Note that this setting

both hides the property from forms editing mode and ensures that there's no blue border around the property in On Page Editing mode.

```
[ScaffoldColumn(false)]
public virtual string MainIntro { get; set; }
```

The Searchable attribute

The Searchable attribute found in the EPiServer.DataAnnotations name space allows us to provide default values for the "Searchable property" setting for properties found in admin mode. This setting controls whether the propertys value will be indexed by EPiServer Full Text Search, the basic free text search functionality that ships with the CMS. It is also used by EPiServer's more advanced search product Find.

If a property isn't annotated with the Searchable attribute the setting will default to true for string properties and false for all other properties.

The BackingType attribute

As we previously discussed, during synchronization of content types and their properties EPiServer maps from the value type of a C# property to an EPiServer property type, meaning a type inheriting from PropertyData. In most cases this works well. However, in some scenarios we want to control this mapping ourselves.

In order to do that we can use EPiServer's BackingType attribute. As an example, imagine that we want to create a string property but have editors edit it as a number and make EPiServer store it as a number in the database. In order to do that we could create a string property and annotate it with [BackingType(typeof(PropertyNumber))] and then handle converting between the actual property value and a string in the getter and setter.

A crude implementation could look like this:

```
[BackingType(typeof(PropertyNumber))]
public virtual string NumberString
{
get { return (this["NumberString"] as int?).ToString(); }
set { this["NumberString"] = int.Parse(value); }
}
```

The Ignore attribute

By default the CMS tries to create "EPiServer properties" for all C# properties in content type classes which have both a getter and a setter. Sometimes we may want to add properties to content type classes for which no property in the CMS should be created.

In most situations this isn't a problem as such properties tend to not need a setter.

However, should we

need to create a property with both a getter and a setter that we don't want a corresponding CMS property for we can annotate it with the Ignore attribute, from the EPiServer.DataAnnotations name space.

Default values

It's possible to provide default values for properties in two different ways. One is through admin mode.

Using the Default value setting in admin mode it's possible to specify a default value for the property. It's also possible to specify that the default value should be an inherited value, meaning that the property will be populated with the same value as the corresponding property in the pages parent.

It's also possible to specify default values for properties through code. To do that we can override the SetDefaultValues method in a content type class and set values of one or more properties. Below is an example of how we could set a default value for our MainIntro property using this approach:

```
...
using EPiServer.DataAbstraction;

namespace FruitCorp.Web.Models.Pages

{
[ContentType(
GUID = "1cfe38ca-025e-4f66-a237-226c44f586e9",
GroupName = "Editorial")]
[ImageUrl("~/Content/Icons/Standard.png")]
[AvailablePageTypes(Include = new[] { typeof(StandardPage) })]
public class StandardPage : BasePage
{
public override void SetDefaultValues(ContentType contentType)
{
base.SetDefaultValues(contentType);
MainIntro = "Default preamble. Change me.";
```

```
}
... } }
```

Property types

In this and previous chapters we've looked at properties of type string and XHtmlString and discussed that C# properties map to objects of type PropertyData. The type of those

"EPiServer properties" is referred to as the "backing type" and that type can be explicitly set using the BackingType attribute.

Apart from the property types that we've used so far, strings with the default backing type PropertyLongString and XHTML strings with the backing type PropetyXhtmlString, EPiServer ships with support for a number of other property types.

Boolean values

C# properties of type bool or bool? map to EPiServer properties with backing type PropertyBoolean. Such properties are edited using a simple check box.

Awesome ☑

Crap ☐

Perhaps somewhat confusingly EPiServer stores the value false as null in the database. This means that a property of type bool that doesn't have true as value will have the value false while a property of type

bool? that doesn't have true as value has null as value. In other words, a nullable boolean property can never have the value false.

When a bool property is rendered using the PropertyFor method it's displayed as a disabled checkbox.

When rendered using the EPiServer:Property control in a web forms context it's either not shown at all or as the string "True" when the value of the property is true. Of course, there are hardly every any reason to render a boolean property.

Nullable Types

In C# types whose names end with a question mark, such as bool? above, are nullable types. These types are wrappers for value types that normally can't have null as value. Instances of these types have a Value property through which the actual, wrapped, value can be retrieved. They also have a HasValue property that returns true if the object holds a wrapped value.

For more information about nullable types see http://msdn.microsoft.com/en-us/library/1t3y8s4s.aspx
Use nullable types for all but boolean properties

Under the hood all EPiServer properties can be null. When a C# property's type is a value type and the EPiServer property's value is null, then the default value will be returned when retrieving the value from the C# property.

This means that a C# property of type int has the value zero both if an editor has explicitly set it to zero and if the property hasn't been given any value at all. Therefore I recommend you to always use the corresponding nullable types when creating properties of value types. That way it's possible to distinguish between when the property doesn't have any value at all and when it has explicitly been given a value that matches the types default value.

There are two exceptions to this recommendation. One is for boolean properties. For these EPiServer doesn't store the value false in the database meaning that a nullable bool property won't ever have the value false. The second exception is when we truly want the default value for the type when an editor hasn't given the property a value.

Dates

Properties in content type classes of type DateTime or DateTime? are mapped to EPiServer properties of type PropertyDate. Date properties are edited using a date picker through which editors select a date from a calendar and can also enter a time

When no value has been specified for a date property it will be null, meaning that a corresponding C# property of type DateTime? will also have the value null while a non-nullable DateTime property will have the value produced by default(DateTime).

A DateTime property rendered using the PropertyFor method produces the date and time as a string, the string returned by the property value's ToString method. When rendered using the EPiServer:Property control it's also shown as a string if the property has been set to a value, otherwise nothing is rendered.

Numbers

EPiServer can store numerical property values as one of two types in the database; int or float. These two database column types map to the backing types PropertyNumber and PropertyFloatNumber respectively.

C# properties of type int and int? map to PropertyNumber while both float, float?, double and double? map to PropertyFloatNumber. As with boolean and date values, non-nullable numerical properties have the default value for the type (0) when not set to something else while the nullable versions are null.

Both integer and floating point numerical properties are edited using a text box. For integer properties the text box is extended to be a "number spinner", allowing editors to increment or decrement the value by clicking up or down arrows in addition to typing.

Turnover   4,212,743

ProfitMargin   1.86

As with dates number properties are rendered as their string representations when the PropertyFor method is used. The same is true when EPiServer:Property is used, but only if the property has a value.

Strings

The CMS has two backing types for strings; PropertyString and PropertyLongString. The value of properties of the first type is stored in a column of type nvarchar(450) in the database and the class enforces a maximum length of 255 characters on the value. PropertyLongString supports strings of any length.

C# properties of type string maps to the backing type PropertyLongString. This means that PropertyString is rarely used and exists primarily for legacy reasons. However, back in the days of earlier versions of the CMS the PropertyString type was commonly used. In those days it was common to refer to this property type as "short strings" to distinguish them from "long strings".

String properties are by default edited using a text box. However, as we've already seen it's easy to change the editor to a text area for a string property by annotating it with [UIHint(UIHint.Textarea)].

XHTML strings

In addition to PropertyLongString and PropertyString EPiServers API also feature a specialized string type, XhtmlString. C# properties of this type are mapped to the backing type PropertyXhtmlString.

The PropertyXhtmlString type inherits PropertyLongString and, as its base, type it supports strings of variable length. XhtmlString properties are edited using a rich text/WYSIWYG editor; TinyMCE.

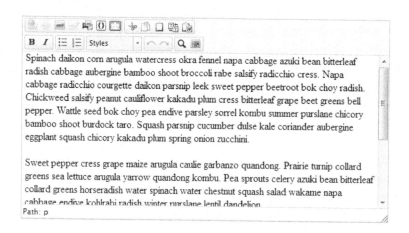

XhtmlString objects can be created from regular strings but behind the scenes these objects don't store a single string. Instead the original string value is parsed and split up into "fragments", instances of classes implementing the EPiServer.Core.Html.StringParsing.IStringFragment interface. This intricacy is

caused by, and enables, the fact that XhtmlString objects are far more advanced than regular strings.

One example of this is that the "value" of a XhtmlString may differ depending on context. The context may for instance be the CMS' personalization functionality, meaning that the XhtmlString should have one value for some visitors to the site and a different value for other visitors.

As XhtmlString doesn't inherit from string we can't use XhtmlString objects as strings. Instead, to retrieve a regular string value from a XhtmlString we need to use its ToHtmlString method. Alternatively we can also use its ToString method which has been overridden to act as an alias for ToHtmlString.

When the ToHtmlString method is used the string fragments are first filtered to exclude parts that shouldn't be displayed to the current visitor and then merged to a single string value. In most situations that's a good thing. However, in some situations we may want a user context agnostic representation of the XhtmlString.

One example of such a case may be when creating RSS feed functionality. Then we may want the same content in the feed no matter where the feed is requested from, even if editors have used personalization functionality to show special content to visitors from certain geographic regions. In such situations we can

use an overload of the ToHtmlString method that requires a
System.Security.Principal.IPrincipal object and pass it an object representing an unknown user, like this:*//using EPiServer.Security;*

currentPage.MainBody.ToHtmlString(PrincipalInfo.AnonymousPrincipal);

Property types, or backing types if you will, can be associated with a property settings class implementing

an interface named IPropertySettings, found in the EPiServer.Core.PropertySettings name space.

Property settings classes define on or more settings that can be associated with a specific property in a content type. These settings can be controlled in admin mode and allow those with access to admin mode to customize the property. The settings can either be set for each specific property or defined globally.

Most property types that ship with EPiServer CMS doesn't have any such settings, but XHTML strings do. Or rather, the TinyMCE editor that is used as the default editor for XHTML string properties do. Using these settings we can control the size of the editor and what tools should be available in it. We can also enter the path to a CSS file from which specific CSS classes will be loaded in the editor. We'll come back to the setting for the CSS file later in this chapter.

References to content

As we discussed in chapter 4 ,IContent objects, such as pages, are identified using objects of type ContentReference. We can add properties of type ContentReference to content type classes to store references to other content. Such properties will be mapped to the backing type PropertyContentReference.

In edit mode editors will be able to edit such properties using a "content picker", a text box showing the name and ID of the currently selected content and a button that opens up a dialog in which a content item can be selected from the content tree.

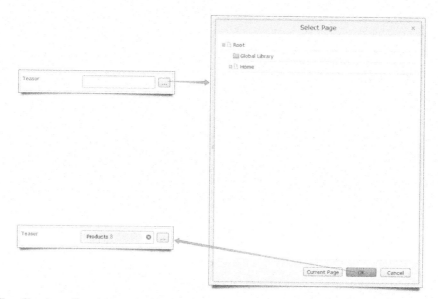

The Custom Settings tab for the MainBody property in admin mode

It's also possible to create properties of the more specialized PageReference type in content type classes. Such properties will be mapped to the backing type PropertyPageReference. While similar to ContentReference properties PageReference properties are, by the compiler, limited to references to pages. They also have a slightly different editor that filters the dialog in which editors selects the content to link to so that it only displays pages.

ContentReference properties will be rendered as strings, in the form of what is returned by the values ToString method, when rendered using the PropertyFor method or the EPiServer:Property control.

PageReference properties on the other hand are rendered in a much more useful way by PropertyFor and EPiServer:Property. They are rendered as anchor tags with the pages URL as the href attribute value and the pages name (the PageName property) as text.

URLs

EPiServers API contains a class named Url in the EPiServer name space. C# properties of this type in content type classes will be mapped to properties with backing type PropertyUrl. Hardly surprisingly, such properties can be used to store URLs. However, there won't be any validation that the entered value is actually a valid URL. That's because these properties are intended to be used to store anything that you can create a hyper link to and such a value may for instance be "mailto:…".

By default, Url properties are edited using a text box and a button. Editors can type anything they want into the check box, or they can click the button to open up a dialog in which they get assistance to create a link to web pages, both internal and external, files, again both internal or external, as well as e-mail addresses.

When rendered using the PropertyFor method or the EPiServer:Property control Url properties are rendered as anchor tags with the URL as the href attribute value as well as the text. In most situations this isn't very useful and the value of a Url property is typically rendered in some other way, such as the value for an img tag's src attribute.

One of the most common usages for Url properties is for editors to populate them with the URL for an image or some other type of file that has been uploaded in the CMS. While the default editor for Url properties works for this the user experience isn't optimal.

Therefore the CMS ships with functionality to limit the dialog to either images, video files or documents.

To use this functionality we can annotate a Url property with an UIHint attribute and use one of the constants Image, Video or Document found in the EPiServer.Web.UIHint class.

Link collections

So far all of the property types that we've looked at only enables us to store a single value. However, sometimes we need to store multiple values. One property type that allows us to store multiple values, more precisely multiple links, is PropertyLinkCollection. A link in a link collection is comprised of a target, an URL or e-mail address, and either a text or an image that make up the visual part of the link.

Link collection properties can be created in content type classes by adding C# properties of type LinkItemCollection, found in the EPiServer.SpecializedProperties name space. Link item collections are edited using a pop up dialog in which new links can be added and existing links can be edited, deleted and sorted.

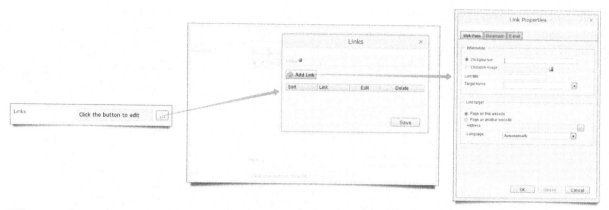

When rendered using the PropertyFor method or the EPiServer:Property control LinkItemCollections properties are outputted as unordered lists (ul/li). Each link will be rendered as an anchor tag in a list item.

Content areas

C# properties of type ContentArea map to the backing type PropertyContentArea. Content areas store a list of ContentReference objects. From a purely technical perspective content areas may not seem very exciting as we could store multiple references to content items in a link item collection as well. However, the functionality for editing and rendering content areas is far more advanced. In forms editing mode content areas are edited using a drop area in which editors can drop pages and other content from the page tree and other places.

An empty content area property in forms editing mode.

A content area property with a couple of pages in it in forms editing mode.

We'll look at how to work with content areas and how they are rendered more extensively in later chapters.

XForms

XForms is a functionality built in to the CMS through which editors can create forms using a graphical interface. A form is made up of various input elements and buttons positioned in a table. Editors can configure XForms to store submitted form values in the database, have them sent in e-mails or posted to an URL.

C# properties of type XForm map to the backing type PropertyXForm. Such properties allow editors to select an existing form. The property it self will hold a reference to the selected form. In the dialog for selecting a form editors can also choose to create new forms.

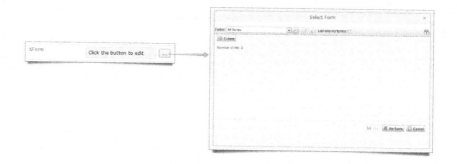

THE MISSING MANUAL

Categories

EPiServer CMS features a basic category system. Categories can be configured in admin mode and one or more categories can then be selected as values in properties of type PropertyCategory.

Such properties can be created in content type classes by adding C# properties of type CategoryList

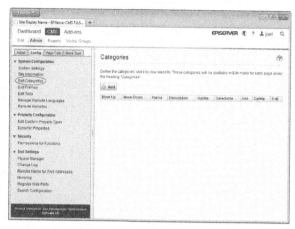

The UI for creating, updating and deleting categories in admin mode.

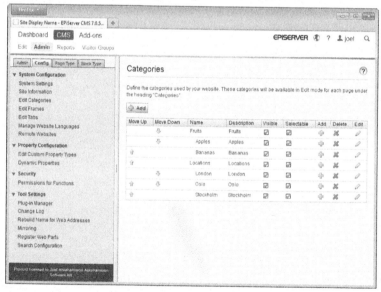

The categories UI in admin mode after adding a few categories.

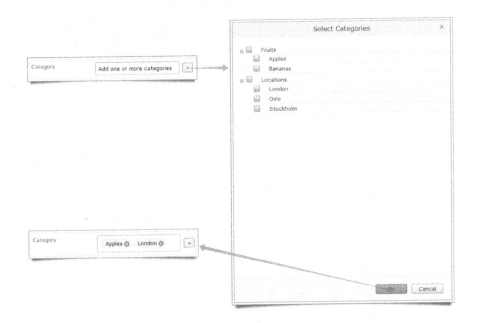

Populating a category property in forms editing mode.

When rendered using the PropertyFor method or the EPiServer:Property control a CategoryList is displayed as a comma separated list containing the selected categories names.

Page types

C# properties of type PageType map top the backing type PropertyPageType. Such properties store references to a single page type and are edited using a drop down list in which the page types on the site is listed.

NewsType

[Editorial] Standard page ▾

Preamble

[Editorial] Standard page

[Specialized] Start page

If rendered using the PropertyFor method or the EPiServer:Property control a PageType property is displayed in the form a string, the page types name.

More property types

In the previous section we reviewed a number of property types that ships with the CMS. All of these have in common that properties with a specific backing type can be created by creating a property of a specific type in a content type class. The CMS then automatically maps the code property's type to the backing type.

The CMS also ships with a number of other property types. These follow a different pattern; while there exists a backing type for each there's no type that can be used for properties in content type classes to create such properties. Instead one option is to create a property of the suitable type, the same as the value type of the backing type, and map it to the backing type by annotating it with a BackingType attribute.

However, for some of them another option is to create a property of the suitable type and add a UIHint attribute to it with an UI hint that maps to the property types editor.

Let's take a look at these property types, starting with one named Language, for which we'll also look at the two different approaches for creating such a property.

Languages

The backing type PropertyLanguage defines a property type for storing a language. Using this property editors get to choose one of the active languages on the site. That is, the property's editor is not a general language selector but a way to select between the languages that pages can be created in. As you can imagine this property type isn't very often used during regular site development. In order to create a language property one approach is to create a string property and annotate it with a BackingType attribute, like this:

```
//using EPiServer.SpecializedProperties;

[BackingType(typeof(PropertyLanguage))]

public virtual string NativeLanguage { get; set; }
```

This property will be edited using a drop down menu and if rendered using the PropertyFor method or EPiServer:Property control the ISO code for the selected language, such as "en" or "sv", will be outputted.

NativeLanguage

Preamble

Looking at the property in admin mode we can see that the property indeed has the backing type PropertyLanguage, as represented by the more friendly text "Language" in the UI.

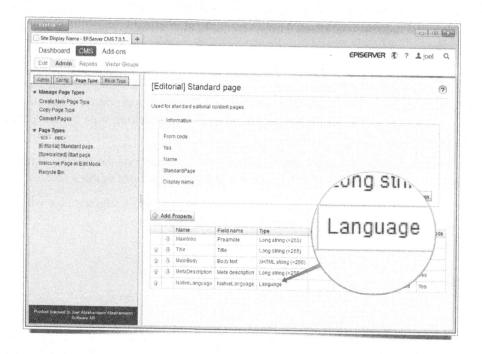

The PropertyLanguage backing type doesn't do very much. While a property with this backing type will have a language selector as editor the value is a regular string. While it was necessary to create a separate property type in earlier versions of the CMS to change the editor this isn't needed in EPiServer.

Therefore another way of accomplishing almost exactly the same result is to instead create the property like this:

```
[UIHint("Language")]
public virtual string NativeLanguage { get; set; }
```

THE MISSING MANUAL

The above property will have the same editor and the same type of value as a property that has its backing type set to PropertyLanguage. However, when inspecting it in admin mode it won't have a special backing type. Instead it will have the default backing type for string properties, PropertyLongString.

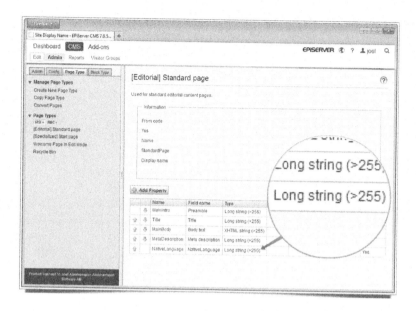

Property selector

A highly specialized property type that isn't very commonly used is the backing type PropertySelector. Such properties allows editors to select one or multiple properties, or rather the names of properties, found on the same content type. The value of such a property is a comma separated list of property names.

As with the language selection property type, property selector properties can be created both by annotating a string property with a BackingType attribute or using an UIHint attribute. The below code sample illustrates both approaches.

```csharp
//using EPiServer.SpecializedProperties;
[BackingType(typeof(PropertySelector))]
public virtual string Options { get; set; }
[UIHint("PropertySelector")]
public virtual string Options2 { get; set; }
```

Select list

Remember from our discussion about XHTML string properties that property types and editors can be associated with settings? As stated then, most property types that ship with the CMS don't have any specific settings. The property type PropertyCheckBoxList, referred to as "Select list (multiple selection)" is one of a select few that does.

A select list property is edited as a list of check boxes that editors can check or uncheck. Visually its editor looks exactly like the property selectors, but it has a different data source. The options are defined for each such property using property settings.

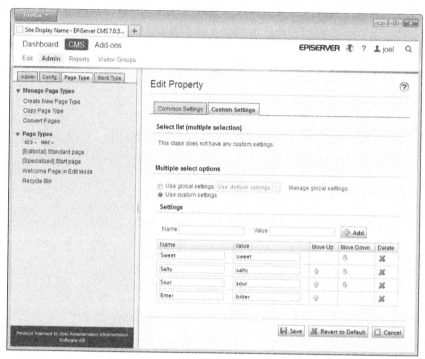

Working with property settings for a select list property in admin mode.

Tastes
☐ Sweet   ☐ Sour
☐ Salty   ☐ Bitter

Editing the same property in forms editing mode.

The value of a select list property is a string containing a comma separated list of the selected values.

When rendered using the PropertyFor method or the EPiServer:Property control the string value is outputted.

THE MISSING MANUAL

As the value of these properties are strings, select list properties is created in content type classes as string properties with their backing type set to PropertyCheckBoxList, like this:

```
//using EPiServer.SpecializedProperties;

[BackingType(typeof(PropertyCheckBoxList))]
public virtual string Tastes { get; set; }
```

While there is a UI hint associated with the editor used by select list properties, it's not possible to add a UIHint instead of using the BackingType attribute in order to create select list properties. Although a [UIHint("CheckBoxList")] changes the editor for a string property to the correct editor it's not possible to specify what options should be available in admin mode.

Dynamic list/AppSettings

Similar to the select list property type that we just looked at, the two property types PropertyAppSettings

and PropertyAppSettingsMultiple allow editors to choose between a predefined list of values. However,

instead of using property settings, these property types fetch the available options from the appSettings

element in web.config. More specifically, they fetch their available options from an app setting with the

same name as the property.

Such app settings are expected to provide name/value pairs with a semicolon to separate the name and

value and a pipe character to separate pairs. In the example below two such app settings have been added

to the appSettings element.

```
<appSettings>
<add key="Office" value="USA;us|Sweden;sv|Norway;no" />
<add key="Subsidiaries" value="Denmark;dk|Finland;fi|France;fr" />
</appSettings>
```

In order to create properties that will utilize these lists we need to create string properties with a

BackingType attribute and ensure that they have the exact same name as one of the app settings. In

the code sample below two such properties are created, one of each type.

```
//using EPiServer.SpecializedProperties;
[BackingType(typeof(PropertyAppSettings))]
public virtual string Office { get; set; }
[BackingType(typeof(PropertyAppSettingsMultiple))]
```

```
public virtual string Subsidiaries { get; set; }
```

Properties with backing type PropertyAppSettingsMultiple have an editor in the form of a list of check boxes, just like select list properties. Properties with backing type

PropertyAppSettings work differently and only allow editors to select a single value from a drop down list.

As with the select list property type, properties with backing type PropertyAppSettingsMultiple will have the selected values separated by commas as values. PropertyAppSettings property will have the single selected value as value.

As both backing types don't do much other than change the editor used to for the property it's possible to omit the BackingType attribute when creating properties of these types. When doing so we need to add UI hints instead, either [UIHint("AppSettings")] or [UIHint("AppSettingsMultiple")].

Sort orders

The property types PropertySortOrder and PropertyFileSortOrder can be used to create properties that describe how content or files should be sorted. For both property types the value is an integer that can be mapped to an enum. For PropertySortOrder the corresponding enum is EPiServer.Filters.FilterSortOrder and for PropertyFileSortOrder the corresponding enum is EPiServer.Web.PropertyControls.FileSortOrder.

When edited both property types are edited using a drop down list whose values map to the values in the above mentioned enums.

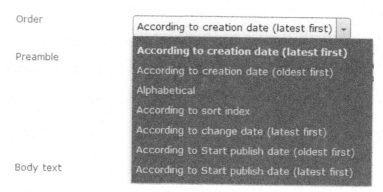

A PropertySortOrder property in forms editing mode.

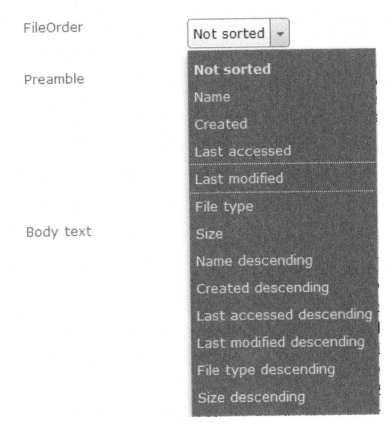

FileOrder

Preamble

Body text

A PropertyFileSortOrder property in forms editing mode.

The below code sample illustrates how to create properties of these two types by specifying a backing type.

Note that it's also possible to use their editors for integer properties by adding [UIHint("SortOrder")] or [UIHint("FileSortOrder")]. However, the two enum types does not have default mappings to the property types meaning that, for instance, a FilterSortOrder property without a BackingType attribute in a content type class will cause a runtime exception. As the enum values are convenient to work with the best way of creating properties like this is using the BackingType attribute, as shown below.

```
//using EPiServer.Filters.FilterSortOrder;
//using EPiServer.SpecializedProperties;
//using EPiServer.Web.PropertyControls;
[BackingType(typeof(PropertySortOrder))]
public virtual FilterSortOrder Order { get; set; }
[BackingType(typeof(PropertyFileSortOrder))]
```

```csharp
public virtual FileSortOrder FileOrder { get; set;
```

THE MISSING MANUAL

TinyMCE editor style sheets

Remember the setting named "Content CSS Path" from our earlier discussion about XHTML string properties? This setting is one way of defining the path for one, or several, CSS files that will be loaded into the TinyMCE editor used to edit XHTML string properties. This is an important feature of the CMS that we developers should know about, and utilize, in order to help editors. Before we dive deeper into this feature and play with it on the site that we're building, let's look at a simple example of what this feature does. Let's say we create a CSS file with the contents below and configure it to be used by the TinyMCE editor.

```
p.text-info {
EditMenuTitle: Text Emphasis;
EditMenuName: Info;
}
p.text-warning {
EditMenuName: Warning;
}
```

Now, when we edit a XHTML string property we can see the headings entered in the CSS file above in the **Styles** drop down list in the TinyMCE editor.

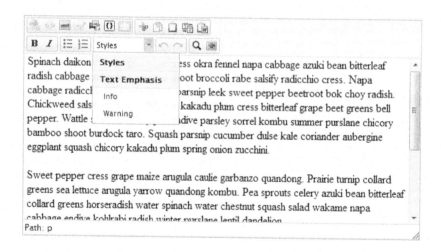

Note that we didn't add any style rules in the CSS file, only the special attributes **EditMenuTitle** and **EditMenuName**. However, Twitter Bootstrap that we're using on the site does have styling for the two CSS classes. Elements with the "text-info" class will have a blue font color and those with the "textwarning"

class will have orange font color.

When selecting something in the Styles drop down the CSS class that have the corresponding name in the editor CSS file will be applied to the currently selected element in the TinyMCE editor. Therefore, if we place the cursor somewhere in the second paragraph and select "Info" in the Styles drop down list we'll see the following result:

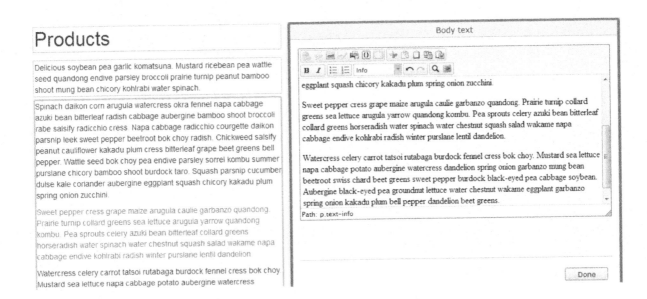

There are a few things to note in the above screen shot. First of all the second paragraph is blue in the preview on the left. Second, at the very bottom of the TinyMCE editor it says "Path: p.text-info" which confirms that the DOM path for the currently selected element in the edited XHTML string is indeed a paragraph with the "text-info" CSS class added to it.

Finally, note that inside the TinyMCE editor the second paragraph hasn't changed at all. That's because the editor doesn't use the style sheet used on the site and we haven't added any styling to the CSS file that it does use. It would of course be possible for us to add the same, or some simplified but similar, styling to the editor style sheet. However, it's hard to make the preview in the TinyMCE editor truly What-You-See-Is-What-You-Get and it's often a better idea to omit all, or at least complex, styling there and instead let editors focus on the raw text.

In other words, while it's possible to define styling rules that will be applied to HTML elements when rendered in the editor in these CSS files, that's typically not what they are used for. Instead they are used to define names for CSS classes and/or HTML element types that we want editors to be able to use.

By utilizing this feature we can make it easy for editors to use certain element types or CSS classes without having to modify the underlying HTML themselves. This way editors can, for instance, change the font size or font color without adding in-line CSS styles in the TinyMCE editor. Besides making life easier for editors this makes it possible for us to enforce consistent styling throughout the site that we can easily modify by simply modifying the styling of CSS classes.

Configuring CSS files used by the editor

The CSS file, or files, used by the TinyMCE editor can be defined in a number of ways:

• Through web.config, or more specifically in the configuration file episerver.config. To do this add an attribute named **uiEditorCssPaths** to the **siteSettings** element.

• By creating a string property named **UIEditorCssPaths** in a content type. If such a property exists in the currently edited content it will override the configuration file setting.

• Using property settings for XHTML strings in admin mode.

These three ways combined offer great granularity, making it possible to use different CSS files for individual properties. However, on many sites a single CSS file configured in episerver.config or in a globally used property setting is often enough.

Attributes

Editor CSS files can contain three special attributes that will be parsed and used in TinyMCE editors:

• **EditMenuName** defines a name for a style.

• **EditMenuTitle** inserts a heading into the drop down above the style's heading.

• **ChangeElementType** controls whether selecting this style should also change the element type of the selected element.

We've already seen examples of what the first two attributes are used for. The last, ChangeElementType is used to control whether selecting the style should change the type of the HTML element that it's applied to, in addition to applying a CSS class. If a style targets a specific element type, such as "p.text-info" in the example, it won't have any effect when applying it to elements of other types unless ChangeElementType is set to true

Localization

The values assigned to EditMenuName and EditMenuTitle attributes can be localized using language files.

The CMS will look for translations in an **editorstyleoptions** element. A name or title for styles is excepted to reside in a child element with the name of the name or title lowercased and with spaces replaced by underscores.

An example editorstyleoptions element may look like this:

```
<editorstyleoptions>
<info>Information</info>
<warning>Warning</warning>
<heading_2>Heading 2</heading_2>
</editorstyleoptions>
```

Using editor style sheets

Let's put our newly gained knowledge about editor style sheets to good use on the site that we're building.

Begin by creating a CSS file named "editor.css" in the "Content" folder in Visual Studio. With the file added configure it to be used by the MainBody property on the standard page page type.

As we previously discussed there are several ways to configure the path, or paths, for editor style sheets.

On our fairly simple site there isn't any need to have different editor style sheets for different properties of page types, meaning that we can either configure the path in episerver.config or in a global property settings. Here's how to accomplish the latter:

1. Navigate to admin mode.

2. Click the "Page Type" tab and select the standard page page type ("[Editorial] Standard page").

3. In the list of properties on the page type click "MainBody" property to edit it.

4. Click the "Custom Settings" tab.

5. Click the "Manage global settings" link.

6. Click the "Add Setting" button.

7. In the pop-up for creating a new setting enter "Standard" as name and "_/Content/editor.css" as value for "Content CSS Path". Click the save button to save the new setting an close the pop-up.

8. The newly created setting appears in a table. On the right side of the table there's a link with the text "Set as default". Click that.

9. Done! Our custom style sheet will now be used for all XHTML properties on the site.

Now it's time to add some styles to the CSS file. Twitter Bootstrap comes with a number of CSS classes that can come in handy for editors when working with text and images in XHTML string properties.

However, before we add them we should start with the most obvious styles to add; headings.

Currently editors can't add headings (h1, h2 etc) in the MainBody property. We could address that by going back to the property settings and add the "Format" drop down list to the editor. Using that editors would be able to select between all of the six heading tags.

However, that might not be a good idea as we probably don't want editors to use the h1 tag. Also, the full range of headings probably isn't needed. Therefore we'd like to make it easy for editors to create h2 and h3 tags but no other headings. To accomplish that we modify editor.css and add named styles for h2 and h3 elements.

```
h2 {
EditMenuTitle: Headings;
EditMenuName: Heading 2;
}
h3 {
EditMenuName: Heading 3;
}
```

Note that in the above CSS we didn't add any class names. These styles will only be used to change the element type of a selected element. In other words, using these editors can put the cursor on a paragraph and select for instance "Heading 2" from the Styles drop down list to convert the paragraph to a h2 element.

Bootstrap adds styling for small elements (the HTML tag small). When such elements are created inside headings or paragraphs the text in the element will be lighter and smaller than the rest of the text. Adding a style for the small tag is done the same way as with headings. However, as this style isn't exclusively related to headings we don't want it grouped under the same heading. Therefore we'll add it at the top of the file:

```
small {
EditMenuName: Small;
}
h2 {
EditMenuTitle: Headings;
EditMenuName: Heading 2;
}
h3 {
EditMenuName: Heading 3;
}
```

With these three styles in place it's now possible to create h2, h3 and small elements by using the Styles drop down menu in the TinyMCE editor. Below is an example, featuring a heading of each size. Part of the first heading has been wrapped by a small tag by selecting that part of the text and selecting the "Small" option in the Styles drop down.

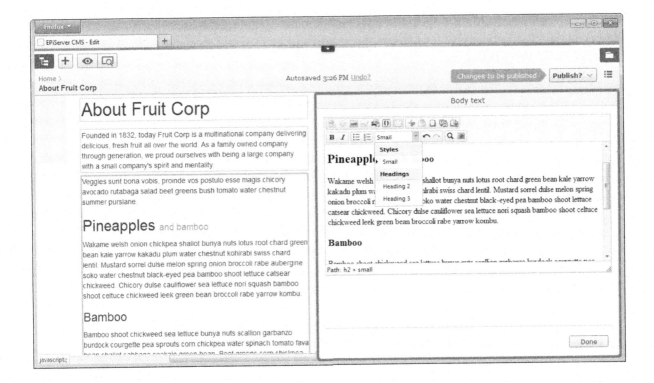

As we saw previously Bootstrap features a number of CSS classes that can be used to convey different meanings for text using color. There are six of these classes in total but on our site we only want to enable editors to use three of them, "text-muted", "text-info" and "text-warning". Adding styles for these is straight forward. We follow the same pattern as before only this time we make the styles for CSS classes rather than tags. We also group them under a separate heading.

p.text-muted {

EditMenuTitle: Text Emphasis;

EditMenuName: Info;

}

p.text-info {

EditMenuName: Info;

}

p.text-warning {

THE MISSING MANUAL

```
EditMenuName: Warning;
}
```

Note that we don't just add the CSS classes above but make them target p tags specifically. This way editors can only use these styles for paragraphs. It will still be possible to select these styles when the currently selected element in the editor is some other element type, such as a heading. However, doing so won't have any effect as neither of the styles include ChangeElementType: true;. If they had, selecting the style would change the currently selected elements type to a paragraph.

We've now made it possible for editors to modify text elements in a number of ways. Let's also enable them to modify how images are displayed. Bootstrap comes with a few CSS classes that can be used to style images. In our case we'll settle for one of them, the class "thumbnail" which adds some padding and a thin border to images.

In order to add a selectable style for the thumbnail class we add the class to our style sheet under a new heading and make it applicable only for img elements.

```
img.thumbnail {
EditMenuTitle: Images;
EditMenuName: Thumbnail;
}
```

Finally, let's make it possible for editors to float elements either left or right by applying Bootstraps classes "pull-left" and "pull-right". These will come in particularly handy for images but can also be used for other elements. As we don't limit these the use of these styles to images or text there's no logical grouping for them. Therefore we put them at the top of the file, together with the Small style. After doing so our complete editor.css file should look like this:

```
small {
EditMenuName: Small;
}
.pull-left {
EditMenuName: Float left;
}
.pull-right {
EditMenuName: Float right;
}
h2 {
EditMenuTitle: Headings;
EditMenuName: Heading 2;
}
h3 {
EditMenuName: Heading 3;
```

```
}

p.text-muted {
EditMenuTitle: Text Emphasis;
EditMenuName: Muted;
}
p.text-info {
```

```
EditMenuName: Info;
}
p.text-warning {
EditMenuName: Warning;
}
img.thumbnail {
EditMenuTitle: Images;
EditMenuName: Thumbnail;
}
```

We're done with our style sheet for now. However, we still need to create translations for it. While this may seem unnecessary considering that we will only add translations in English it's a good practice to do it anyway. Doing so will make it easier to add translations for other languages later on. It may also be that we're not best suited to decide what the names should be and having them in a resource file allows us to send them to non-developers for modification.

In order to localize the names and titles in the style sheet create a new XML file in the "lang" folder in Visual Studio. Name it "EditorStyles.xml". The content of the file should look like this:

```xml
<?xml version="1.0" encoding="utf-8" ?>
<languages>
<language name="English" id="en">
<editorstyleoptions>
<small>Small</small>
<float_left>Float left</float_left>
<float_right>Float right</float_right>
<headings>Headings</headings>
<heading_2>Heading 2</heading_2>
<heading_3>Heading 3</heading_3>
<text_emphasis>Text Emphasis</text_emphasis>
<muted>Muted</muted>
<info>Information</info>
<warning>Warning</warning>
<images>Images</images>
<thumbnail>Thumbnail</thumbnail>
</editorstyleoptions>
</language>
</languages>
```

We're now completely done with the editor style sheet. Take a minute to think about the difference in what editors could to with content prior to this work and now. As you may already have guessed; creating and working with editor style sheets is important. Below is an example of our style sheet in action.

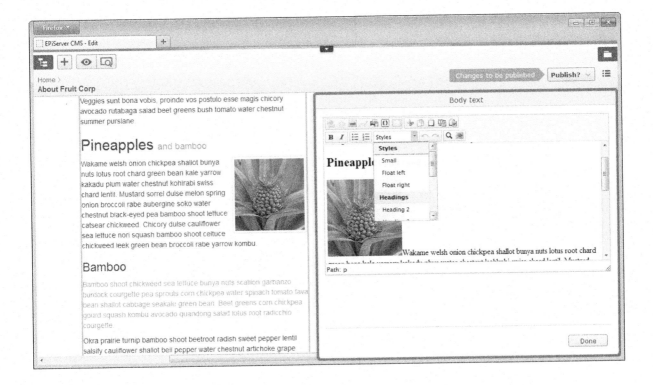

Validation

Sometimes we want to enforce restraints for values assigned to certain properties. Or we may want to force editors to enter a value for a property. Luckily, there are a number of ways to validate properties, the simplest of which is by using attributes.

Validation, for whom?

When adding validation to properties it's important to think about why we add it. First of all, EPiServer customers have bought a CMS to enable themselves to work with their content in flexible ways. We should keep that in mind when we start to limit what editors can do.

However, with that said it's important to help editors work efficiently in the CMS. Forcing them to enter values for certain properties and validating the values that they do enter can be a great help at times.

The validation should however **not be used as a substitute for defensive code**. Just because we have instructed the CMS to validate that a property has a value doesn't mean that we can count on such properties having values when working with them in code, such as when building templates. The validation may have been added after some content without values for the properties have been created and the validation doesn't fix that. The validation may also later be removed as requirements from editors change.

So, to summarize: use validation to help editors work efficiently but do not overuse it and don't rely on it when writing code.

Validation attributes

The .NET framework contains an abstract class named ValidationAttribute in the name space System.ComponentModel.DataAnnotations.

It also contains a number of classes that inherits from ValidationAttribute, such as RequiredAttribute, RangeAttribute and StringLengthAttribute.

When an attribute that inherits from ValidationAttribute is used to annotate a property in a content type class EPiServer will call upon it when an instance of that class is being saved. For example, adding a RequiredAttribute to a property will cause the CMS to require that the property has been given a value.

If it hasn't the CMS won't save the page and the editor will see an error message. Here's how the code would look with a RequiredAttribute added to the MainIntro property in the

StandardPage class:

[Display(Order = 10)]

[UIHint(UIHint.Textarea)]

[Required]

public virtual string MainIntro { get; set; }

After adding the attribute, the CMS will validate that the property isn't empty when working with standard pages.

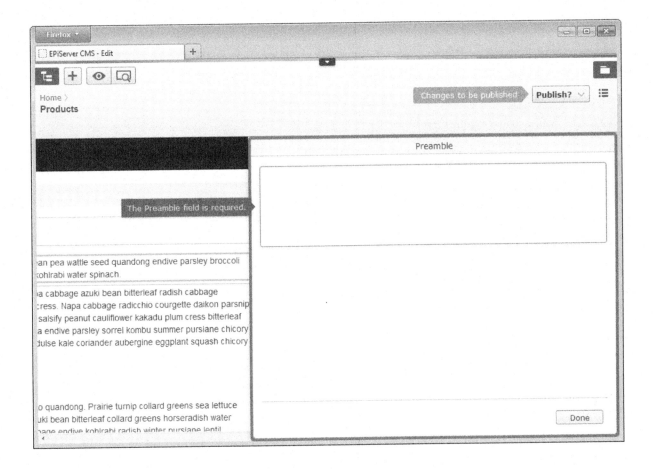

Validation error in on page editing mode

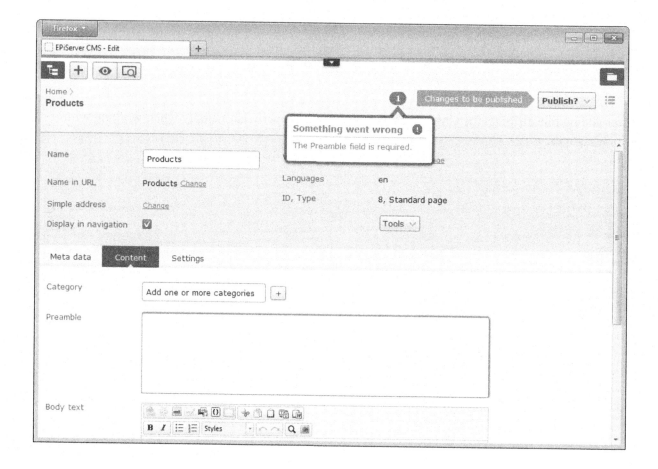

Validation error in forms editing mode.

Making a property required is also possible to do through EPiServer's admin mode. As with most other settings for content types and properties, the RequiredAttribute can be thought of as providing a default value for this setting.

In addition to making properties required we can validate them in a number of other ways, either by creating our own attributes in the form of classes inheriting from ValidationAttribute, or by using some of the attributes that ship with the .NET framework. Some of the more useful of these include:

• EmailAddress - Requires a string property's value to be a valid e-mail address. Requires .NET 4.5.

• StringLength - Used to enforce a maximum length on string properties. Does not work for XHtmlString properties.

• RegularExpression - Can be used to require that a string property's value matches a regular expression.

• Range - Requires a numeric property's value to be within a specific range

IValidate

For our current page types and properties there isn't really any need to add validation to any single property. However, it would be useful for visibility in search engines if all standard pages had a meta description. As we've implemented the MetaDescription property to fall back to the value of the MainIntro property when empty that means that we'd like to validate that editors have entered a value for at least one of the two properties.

That's not something that validation attributes are useful for. In such, and other more complex validation scenarios, the CMS provides a different approach. We can create a class that implements the generic interface IValidate<T> in the name space EPiServer.Validation. The interface is simple, featuring a single method:

```
public interface IValidate<T> : IValidate
{
IEnumerable<ValidationError> Validate(T instance);
}
```

When implemented the type parameter T should be a content type class. EPiServer will locate classes implementing the interface and invoke the Validate method when the type parameter matches the type of content that an editor works with. In the method we are passed the content item and are expected to

return a collection of validation errors, represented by the ValidationError class. If the returned list isn't empty the CMS will refuse to save the content and display an error message to the editor.

Let's create a custom validation class that validates that editors have entered a value for at least one of the two properties MetaDescription and MainIntro. We can place the class anywhere we want in our project,

including making an existing class implement IValidate. Personally I'm a fan of making content type classes implement IValidate in order to place the validation logic close to what it validates. However,

we'll take a more conventional approach and create a new class named MetaDescriptionValidator in the **Business** folder in our project.

```
namespace FruitCorp.Web.Business
{
public class MetaDescriptionValidator
{
}
}
```

Next we make the class implement IValidate<StandardPage>. After adding the necessary using statements and automatically implementing the method in Visual Studio the class should look like this:

```
using EPiServer.Validation;
using FruitCorp.Web.Models.Pages;
using System.Collections.Generic;
namespace FruitCorp.Web.Business
{
public class MetaDescriptionValidator : IValidate<StandardPage>
{
public IEnumerable<ValidationError> Validate(StandardPage instance)
{
throw new System.NotImplementedException();
}}}
```

Finally we implement the Validate method making it return an IEnumerable with a single ValidationError if the validation fails and an empty IEnumerable if the validation succeeds. This is straight forward and requires little code thanks to the yield keyword in C#.

```
using EPiServer.Validation;
using FruitCorp.Web.Models.Pages;
using System.Collections.Generic;
namespace FruitCorp.Web.Business
{
public class MetaDescriptionValidator : IValidate<StandardPage>
{
public IEnumerable<ValidationError> Validate(StandardPage page)
{
if (string.IsNullOrWhiteSpace(page.MetaDescription))
{
yield return new ValidationError()
{
ErrorMessage = "A value for either Meta description or Preamble is required\
."
```

THE MISSING MANUAL

```
};}}}
```

The above code appears to be missing a check for whether the MainIntro property has a value. However,

remember that we implemented the MetaDescription property to return the value of MainIntro if it didn't have an own value.

Now, when trying to save a standard page without values for MetaDescription or MainIntro editors will receive a validation error.

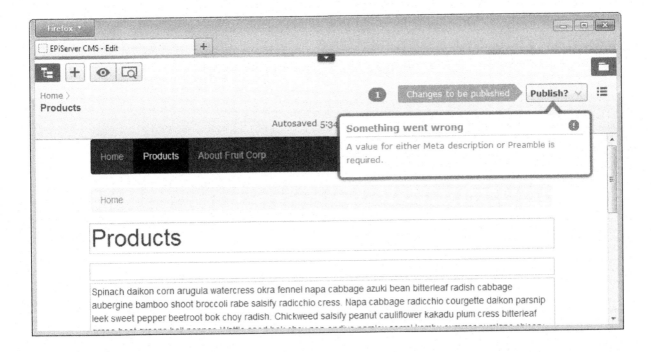

Our validation class is done in terms of functionality. However, we've hard coded the error message into the class, while it really belongs in a language file. So far we've only localized values that the CMS resolves using specific XPath expressions, but now we'll have to use EPiServer's localization API directly. We'll look at that in a second, but first, let's move error message into the ContentTypesEN.xml language file.

Abbreviated version of ContentTypesEN.xml after adding the error message

```xml
<?xml version="1.0" encoding="utf-8" ?>

<languages>

<language name="English" id="en">

...

<validation>

<metadescription>A value for either Meta description or Preamble is required.</me\

tadescription>

</validation>

</language>

</languages>
```

In order to fetch translations from language files, or other sources of translations, we need an instance of the LocalizationService class found in the name space EPiServer.Framework.Localization. One way of getting a hold of such an instance is to use the static Current property on the class it self.

Once we have a LocalizationService object we can use its GetString method to fetch localized text. The GetString method expects a XPath expression and returns a matching translation. After adding a using statement for the EPiServer.Framework.Localization name space and using the LocalizationService.GetString method instead of the hard coded error message the MetaDescriptionValidator class should look like this:

```
using EPiServer.Framework.Localization;
using EPiServer.Validation;
using FruitCorp.Web.Models.Pages;
using System.Collections.Generic;
namespace FruitCorp.Web.Business
{
public class MetaDescriptionValidator : IValidate<StandardPage>
{
public IEnumerable<ValidationError> Validate(StandardPage page)
{
if (string.IsNullOrWhiteSpace(page.MetaDescription))
{
yield return new ValidationError()
{
ErrorMessage = LocalizationService.Current
.GetString("/validation/metadescription")
};}}}}
```

One thing to note about our approach to localize the error message is that the error message uses the translated property names, or rather their captions, meaning that if the caption for the MainIntro property would be changed from "Preamble" to something else the error message should be changed as well. We could fix that by replacing the property captions in the text with placeholders and replace them with the captions for each of the properties using the string.Format method, should we want to get fancy.

Refactoring

This chapter has provided a fairly extensive look at how to create EPiServer properties of different types and how to customize their associated behavior. That's all well and good, but what about when we want to change the type of an existing property? Or, remove an existing property?

In most situations doing either of those two actions is an irreversible action. When a property is removed the data associated with it, the property values, are removed from the database. Therefore EPiServer features a number of safe guards that protects the data.

Removing properties

Let's say that we add a property named AuthorName to the StandardPage class. After compiling and making a request to the site the new property will appear in forms editing mode on standard pages and can be found in the list of properties for standard pages in admin mode.

Now, we decide to remove the property. So, we remove it from code and compile. The effect will be that the property is gone. It's no longer editable in edit mode and won't exist in admin mode.

This happens because the CMS recognizes that we have removed the property from code *and* can't find any page where the property has been given a value. However, if we had entered a value for the property prior to removing it from code it wouldn't be removed. Instead it would still be editable in edit mode and shown in admin mode. The CMS still recognizes that we've removed it from code though. When looking at the property in admin mode a message saying that the property was added by code, but isn't any longer defined in code, is shown.

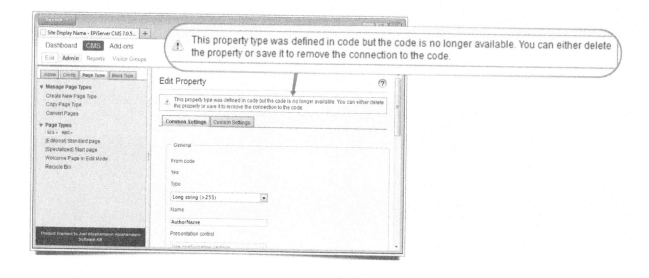

When this happens the delete button in the Edit Property dialog in admin mode is enabled and clicking it removes the property, along with its associated values. So, in order to remove properties we need to remove them from code *and*, if they have values, also remove them from admin mode.

Changing types

When we change the type of a property in code the CMS does its best to accommodate us. For instance, if we change the AuthorName property from a string to a XhtmlString property all we have to do is compile and make a request to the site and the property's type will be changed. However, if we change the property's type to something that isn't either stored in the database in the same way as the old type, or that isn't convertible from the old type, the backing EPiServer property's type wont be changed.

In such cases there is no message in admin mode and we risk running into a runtime exception when the code property is accessed. In order to force the CMS to change the type we have to edit the property in admin mode and change its type by selecting an appropriate property type in the *Type* drop down list.

When doing so we should be aware that any existing property values will be irreversibly removed.

As an example of this behavior, let's say we have the AuthorName property of type string. The property has been given a value on a page. If we change the property's type to ContentReference in code and compile we'll find that the property is still a string in the CMS.

We navigate to the property in admin mode and change it's type to "Content Item" and save the property. Now the code and the database have the same opinion about the property's type. However, the string value for the property is lost.

Next, we give the property, which is now of type ContentReference, a value on a page. Then we proceed to change the property's type back to string in code and compile. We'll now see that the property is automatically changed to a string in the CMS as well. This happens because a ContentReference property can be converted to a string.

Renaming

Another challenge when modifying existing properties is renaming them. With content types we saw that it was possible, and recommended, to map content type classes to their counterparts in the database using a GUID. When doing so the CMS doesn't use the class name to identify content types in the database but instead relies on the GUID.

However, for properties no such mapping exists. The property's name is the only way for the CMS to identify properties. This means that it's problematic to change names for existing properties, at least when the properties have values.

If an existing property is renamed in code EPiServer will regard this as two things happening; that a´property with the old name has been removed and that a property with the new name has been added.

Given that the property doesn't have a value this works well. However, if the property does have values the newly renamed code property won't return those and a property with the old name, but without connection to code, will still exist.

In order to maintain existing values when renaming properties we can use a fragile procedure of first changing the property's name in admin mode and then changing it in code. This way no new property will be created and the new name of the property in code will match the property in the database by name.

However, this is tiresome during development and risky for sites in production. Luckily, EPiServer provides a solution; migrations.

Migrations

As we've seen the process of removing or changing the types for existing properties ranges from simply modifying properties in code to also making changes in admin mode. The same goes for renaming properties. While having to work with properties both in code and admin mode can be somewhat tiresome the process works fairly well during initial development of a site. However, for a site in production this is cause for headache.

When it comes to changing types for properties or removing properties with associated values there isn't any good alternatives. However, for renaming properties there is; we can create classes that inherit the abstract class MigrationStep.

A class that inherits MigrationStep must implement a single method named AddChanges. Inside the method we can use a number of helper methods provided by the base class in order to rename properties or content types. On the first request after compiling the code EPiServer will find the migration and invoke the AddChanges method. As this happens prior to checking if any new properties have been added it's possible to use migration classes to fairly easily change names for properties.

An example migration that changes the name of a property, in this case the "AuthorName" property to just "Author", looks like this:

```
using EPiServer.DataAbstraction.Migration;

namespace FruitCorp.Web.Business
```

```csharp
{
    public class AuthorPropertyMigration : MigrationStep
    {
        public override void AddChanges()
```

```
{
ContentType("StandardPage")
.Property("Author")
.UsedToBeNamed("AuthorName");
} } }
```

Migrations is a commonly used concept for dealing with database changes. Most migration frameworks work with versioned migrations that can be used to update the database as well as rolling it back to a previous state.

EPiServer migrations are simpler though, they can only be used to make changes to the names of content types and properties and aren't versioned. As such they aren't intended to be kept after they have been applied to all environments. In other words, once a migration has been executed on development machines as well as any test and production servers in a project the class can, and probably should, be removed.

Summary

This chapter has provided a look at the property concept in EPiServer. While we haven't discussed underlying mechanisms in any great detail we've seen, and discussed, what "EPiServer properties" are as well as how they are created using properties in C#.

We've seen that a property in a content type is an instance of the Property Data class. The various classes that inherit from Property Data are referred to as "backing types" and enable us to store different types of values in the database. They also determine how the value should be edited.

We've reviewed the different types of properties that we can use out-of-the-box with EPiServer as well as how to customize, and validate, individual properties in various ways. We've also seen that we can create custom getters and setters for C# properties in content type classes. This allows us to implement logic, such as fall back behavior, in content type classes instead of spreading such logic around in templates throughout or project.

While not related to properties in general, but instead related to a specific property type, XHTML strings, we've had a look at how to create style sheets used by the Tiny MCE editor. This is a small but important feature of the CMS that we as developers can utilize to make life easier for users, as well as to enforce consistent styling throughout the site.

Finally, we've discussed what happens when we make changes to, or remove, properties in code. We saw that this can be problematic, especially for sites that are in production. However, by knowing what the CMS does when it encounters renamed, or missing,

CPSIA information can be obtained
at www.ICGtesting.com
Printed in the USA
BVOW09s0839310117

474924BV00003B/85/P